Advanced Materials and Techniques for Radiation Dosimetry

For a listing of related *Artech House* titles,
turn to the back of this book.

Advanced Materials and Techniques for Radiation Dosimetry

Khalil Arshak
Olga Korostynska
Editors

ARTECH HOUSE

BOSTON | LONDON
artechhouse.com

Library of Congress Cataloging-in-Publication Data
A catalog record for this book is available from the U.S. Library of Congress.

British Library Cataloguing in Publication Data
Advanced materials and techniques for radiation dosimetry.
 —(Artech House sensors library)
 1. Radiation dosimetry
 I. Arshak, Khalil II. Korostynska, Olga
 539.7'7

ISBN-10: 1-58053-340-x

Cover design by Bob Pike

© 2006 ARTECH HOUSE, INC.
685 Canton Street
Norwood, MA 02062

All rights reserved. Printed and bound in the United States of America. No part of this book may be reproduced or utilized in any form or by any means, electronic or mechanical, including photocopying, recording, or by any information storage and retrieval system, without permission in writing from the publisher.
  All terms mentioned in this book that are known to be trademarks or service marks have been appropriately capitalized. Artech House cannot attest to the accuracy of this information. Use of a term in this book should not be regarded as affecting the validity of any trademark or service mark.

International Standard Book Number: 1-58053-340-x
ISBN 10: 1-58053-340-x
ISBN 13: 978-1-58053-340-9

10 9 8 7 6 5 4 3 2 1

In memory of my parents—Ibraham and Mary Arshak and to my wife Arous Arshak and my children, Annie, Raffie, and Addis
—*Khalil Arshak*

To my parents, Tamara and Stanislav, and my brother Andrew
—*Olga Korostynska*

Contents

	Preface	xi
1	**Introduction**	**1**
	References	8
2	**Radiation Dosimetry: Background and Principles**	**11**
2.1	Review of Radiation Types	11
2.1.1	Ionizing Radiation	11
2.1.2	Stopping Power and Linear Energy Transfer	14
2.1.3	Radiation Units	24
2.2	Biological Effects of Radiation	25
2.2.1	Types of Radiation Effects	25
2.2.2	Relative Biological Effectiveness	28
2.3	Basic Principles of Radiation Protection	31
2.4	Dosimetry for Industrial Radiation Processing	33
2.4.1	Medical Devices Sterilization	33
2.4.2	Food Irradiation	34
2.4.3	Modification of Polymers	35
2.5	Medical Use of Ionizing Radiation	37
2.5.1	In Vivo Dosimetry	38

2.6	Uncontrolled Radioactive Releases	39
2.6.1	Nuclear Accidents	40
2.6.2	Radioecology After Nuclear Accidents	41
2.6.3	Further Examples of Nuclear Contamination	45
2.7	Review of the Principles and Materials in Radiation Dosimetry	46
2.7.1	Dosimetry	46
2.7.2	Classification and Calibration of Dosimeters	48
2.7.3	Gas-Filled Detectors	51
2.7.4	Scintillation Counters	57
2.7.5	Chemical Dosimeters	61
2.7.6	Radiochromic Dye Films	61
2.7.7	TLDs	62
2.7.8	Radiation Dosimetry Using MOS Devices	70
2.7.9	Interaction of Ionizing Radiation with Semiconductor Detectors	71
2.7.10	Response of Diodes to External Exposure	75
2.7.11	Other Radiation-Sensitive Materials of Interest	79
	References	84

3 Effect of Radiation on Optical and Electrical Properties of Materials — 91

3.1	Introduction	91
3.2	Optical Absorption	92
3.3	Amorphous Films	94
3.4	Absorption Spectra of Amorphous Solids	96
3.5	Metal-Semiconductor Contacts	99
3.5.1	Ohmic Contact	99
3.5.2	Neutral Contact	99
3.5.3	Blocking Contact	100
3.6	Conduction Mechanisms in Amorphous Materials	101
3.6.1	Schottky Versus Poole-Frenkel Conduction Mechanism	104
3.7	Radiation Damage in Crystalline Structures	105

3.8	Radiation-Induced Defects in Oxide Materials	106
3.9	Radiation Effects in Polymers	108
3.10	Radiation-Induced Degradation Processes in Device Parameters	112
	References	113
4	**Gamma Radiation Dosimetry Using Metal Oxides and Metal Phthalocyanines**	**115**
4.1	Thin and Thick Film Technologies	115
4.1.1	Thin Film Technologies	115
4.1.2	Thick Film Technology	116
4.2	Thin Films as Radiation Sensors	117
4.2.1	Metal Oxides Thin Films	118
4.2.2	Metal-Substituted Phthalocyanines Thin Films	132
4.3	Thick Films as Radiation Sensors	137
4.3.1	Effects of γ-Rays on the Optical and Electrical Properties of Metal Oxide Thick Films	137
4.3.2	Effects of γ-Rays on the Optical Density of MePc Thick Films	145
4.3.3	MOS Dosimeter Using Bismuth Oxide (Bi_2O_3) and Copper Phthalocyanine (CuPc) Polymer Thick Films	150
4.4	Conclusion	153
	References	154
5	**Sensor Arrays, Radiation Nose Concept, and Pattern Recognition**	**159**
5.1	Sensor Arrays	159
5.1.1	In_2O_3/SiO Sensor Array	159
5.1.2	PolyVinyliDene Fluoride Films of Different Thicknesses	161
5.1.3	Multiple Sensor Materials Incorporated into an Array	162
5.2	Dosimetry of Mixed Radiations	162
5.2.1	Electronic Nose Systems	163
5.3	Radiation Nose Concept: Overview of Pattern Recognition	165

5.3.1	Radiation Nose Concept	165
5.3.2	Overview of Pattern Recognition	168
5.3.3	Feature Extraction/Selection	171
5.4	Classification and Validation	172
5.4.1	Classification	172
5.4.2	Principal Component Analysis	172
5.4.3	Linear Discriminant Analysis	175
5.4.4	Validation	175
5.5	Radiation Nose Using a Compaq iPAQ Designed for Resistive Sensors	176
5.6	Portable Real-Time Gamma Radiation Dosimetry System Using MgO and CeO_2 Thick Film Capacitors	179
5.6.1	Change of Capacitance with Dose	179
5.6.2	Circuit	180
5.6.3	Software	181
	References	186
6	**Conclusions and Future Trends**	**189**
6.1	Necessity of Radiation Dosimetry	189
6.2	The Choice of Detector	189
6.3	Standards and Requirements	190
6.4	Future Trends	191
	References	191
	Acronyms	**193**
	Appendix	**199**
	About the Authors	**201**
	Index	**203**

Preface

The purpose of this book is to draw attention to the modern body of knowledge of radiation sensors with a focus on the wide range of gamma radiation dosimeters, both commercially available and under development. The book provides knowledge of most common and emerging dosimetry methods and materials used for radiation measurement. A research approach to discuss the properties of materials under the influence of ionizing radiation is expected to be of an interest to both students and engineers. General concepts and theoretical considerations would target the attention of both radiation sensors designers as well as general audience members, who may wish to increase their level of knowledge in this area. The readers level assumes an elementary background in radioactivity and radiation as well as a basic knowledge on electrical, optical, and structural properties of materials and device operation. It is hoped that the reader of this book will gain a better understanding of the main principles of radiation physics and acquire advanced knowledge of radiation-induced changes in the materials properties, which would be successfully used in the future to design novel affordable dosimetric devices. The theoretical part of this book would be of interest to most undergraduate and postgraduate students, as it contains basic principles of radiation physics and general radiation protection information. Lecturers and students in radiation physics, health physics, radiation biology, nuclear chemistry, and so forth would consider this book a useful up-to-date guide to various radiation sensors.

The book welcomes the reader in the world of radiation and discusses its types and sources, the biological effect of the radiation, and its interaction with matter. Advantages and disadvantages of radiation usage are discussed in terms of its positive effects in medicine and industry as well as its negative consequences, such as nuclear accidents. Classification, operation principles, and examples of commercially available dosimeters for various applications are presented. A review of the newest findings on gamma radiation dosimetry using metal oxides and polymers, such as metal-substituted phthalocyanines, is given. A novel concept of radiation nose is presented for the first time. In the radiation nose system, an array of sensors is used to monitor various ranges and types of ionizing radiation, therefore increasing the working dose range and the sensitivity of the overall system. Based on a set of the output readings from the sensors, a specific pattern recognition (PR) algorithm is applied, so that the radiation nose system accurately determines radiation dose and type. The overview of PR algorithms along with feature extraction/selection, classification, and validation is given.

1

Introduction

The presence of natural radiation in the environment arises from cosmic and terrestrial sources. Earth is being constantly bombarded with cosmic rays (high-energy protons and charged nuclei) of solar and galactic origin. These cosmic rays cause direct exposure, but they also give rise to secondary radioactivity from their interaction with stable elements in the upper atmosphere, forming radionuclides that can give rise to human exposure by inhalation or by ingestion after their uptake by plants. These radionuclides include Tritium (^{3}H), Beryllium-7 (^{7}Be), Carbon-14 (^{14}C), and Sodium-22 (^{22}Na) [1]. The most significant cosmogenic radionuclide is ^{14}C, which is then taken up by plants and becomes incorporated into human foodstuffs.

Most elements with atomic number Z between 1 and 92 (Uranium) exist naturally, and those from 93 to 106 have been produced artificially. Promethium (Z = 61) is an exception, as it doesn't occur naturally but is a fission product of uranium, thorium, and plutonium. All elements with Z > 82 (lead) are radioactive and undergo nuclear rearrangements with emission of subatomic particles and gamma radiation until a stable configuration is reached. Unstable nuclei may occur naturally or form as a result of human activities. Instability of atomic nuclei may arise from excess energy within the nucleus or an imbalance between the number of protons and neutrons. Disintegration of an unstable nucleus may occur spontaneously or follow interaction with a nuclear particle. Forms of radiation that may be emitted from nuclei include alpha particles (identical with ^{4}He nuclei), beta particles (electrons), gamma radiation (very short wavelength electromagnetic radiation), or neutrons. Alpha particles often carry the most energy, but—because of their double positive charge—they interact strongly with matter and are readily stopped by, for example, a sheet of paper. Beta particles, with a single negative charge, are usually more penetrating than alpha particles and can be stopped by a thin sheet of metal. The neutron is uncharged and only interacts with matter when passing close to the atomic nuclei, so that neutron radiation is very penetrating. Gamma radiation is the most penetrating of all and can require many centimeters of heavy metal or several meters of concrete to stop when at high energy (short wavelength).

We contain many radioactive materials ourselves, such as a radioactive form of potassium (^{40}K) that was created when the earth was first formed. The concentration of ^{40}K in the body is held relatively constant by metabolic processes, and its annual dose is estimated to be 165 μSv [1]. Common building materials, such as marble and granite, contain measurable amounts of natural radon, uranium, and thorium isotopes.

The average annual effective dose from natural radiation that the U.K. population receives is 2.23 mSv, and about half of this is from Radon (^{222}Rn) exposure indoors [1]. ^{226}Ra is present in the ground and decays to ^{222}Rn gas, which gives rise to an inhalation dose. The average annual dose from artificial radiation is a little over 0.42 mSv and is mainly due to the use of X-rays in medical procedures. The contribution from this source has increased by about 10% in recent years due to the large number of computed tomography procedures. The overall average annual dose is therefore almost 2.65 mSv.

Government standards for radiation protection in the United States are established by the National Council on Radiation Protection and Measurement (NCRP) and its international counterpart, the International Commission on Radiological Protection (ICRP). Both of these organizations offer recommendations for the maximum permissible dose of radiation to which people should be exposed—that is, 1 mSv per year for general public and 20–50 mSv per year for radiation workers [2]. In contrast to the dosage measurements for external radiation, internal doses cannot be measured directly: they must be inferred from the measurement of quantities such as body activity content, excretion rates, or airborne concentrations of radioactive materials. Furthermore, the assessment of exposures due to intake depends critically upon knowledge of the biokinetics of the radionuclides [3].

Radiation processing is an expanding technology with numerous applications in, for example, health care products sterilization, sewage and hospital waste treatment, polymer modification, and food processing. The effectiveness of the irradiation process depends on the proper application of dose and its measurement. The required absorbed dose range would depend on both the product and the desired effect. Adequate dosimetry with proper statistical controls and documentation is the key factor of the quality control process, which is necessary to assure the products are properly treated.

Real-time radiation detectors are essential equipment for emergency personnel, who may have to respond to unknown accidents, incidents, or terrorist attacks, any of which could involve radioactive material. More and more ordinary citizens are interested in personal radiation protection as well. Reasons include lost sources, nuclear industrial accidents, nuclear or radiological terrorism, and the possibility of nuclear weapons being used in a war. People want to have the ability to measure it for themselves, and they want to be notified when the radiation levels are increased. An ordinary handheld radiation meter is usually not available when it is really needed. The policeman responding to an emergency situation might not be aware that radioactive material is involved, and he will not even think about bringing a Geiger counter. The person who picks up a lost radioactive source, without knowing what it is, or who walks into a radiography shoot does not carry a meter. Unfortunately, a situation may occur in which a strong radioactive source could be planted in a crowded location, such as a subway, stadium, or cinema. It is hardly possible to maintain real-time radiation monitoring of all public places, but one must minimize possible consequences of radiation overexposure with timely alert, given by dosimeter. An increasing number of workers who use personal dosimeters—apart from previously mentioned emergency medical personnel, police, and military forces—include customs and border patrols, security officers in banks, government laboratories and agencies, and fire departments. Radiation awareness has increased dramatically over the last decade. With the anticipated general public anxiety, perhaps even panic, accompanying any future developing nuclear emergency or even the transportation of nuclear waste, it will be very reassuring to know with confidence whether or not there is any danger of radiation presence. In all cases, knowing exactly what the radioactivity is, where you are standing, will always enable better-informed decisions, along with the ability to take correct protective action to minimize any future radiation exposure for you and your family.

Chapter 2 presents some theoretical background on the basics of radiation dosimetry and measurement units. It discusses various types of ionizing radiation, interaction of radiation with matter, radiation-induced biological effects, and the benefits and disadvantages of its industrial usage. Basic principles of radiation protection are given, so one can be alert in case of emergency.

Radiation processing forms an important component of several industrial sectors (e.g., the polymer and medical device industries). One of the principal advantages of ionizing radiation as an industrial tool is the ability to achieve precise chemical and biological effects by the delivery of known doses of radiation. Currently, there are three main applications for industrial radiation processing: the sterilization of medical devices, the treatment of foodstuffs, and the modification of polymers.

The requirements of radiation sensors include high sensitivity and linear performance over the intended energy range, real-time response, low noise, and acceptable reliability under the exposure conditions [4]. Due to the large number of different applications, there are many variants of the radiation sensors, which use different materials, geometric arrangements, and physical-detection techniques. The choice of a particular detector type depends on application, ionizing radiation energy range, resolution, and efficiency requirements. Additional considerations include the performance, timing, and environmental suitability of the detector and, of course, the price. Examples of commercially available personal radiation dosimeters include a pen-type dosimeter, film or ring badge, thermoluminescent dosimeters, and various electronic pocket dosimeters, such as personal electronic dosimetry cards, and so on. If used for personnel dosimetry, the product preferably should be small and light enough to be worn as a badge, watch, belt clip, or similar device and be manufactured easily at an industrial level with low cost. In some applications, radiation sensors have to operate in harsh conditions, such as high temperatures, aggressive gases, and the presence of humidity. The list of companies (including their Web addresses) that sell or manufacture various radiation detection equipments can be found in the Appendix.

The process of using radiation to sterilize medical products, such as drip bags, syringes, bottles, and implants, is completely automated. Once initial characterization has been performed for a given product, total dose exposure is exclusively controlled by exposure time. Dosimeters to monitor the radiation are usually made of cast polymethylmethacrylate (PMMA) sheets. They cost approximately $1 each and are used in amount of about 50,000 per year by a small-scale enterprise. Their optical properties change under exposure to gamma radiation, and this gives an indirect indication of the total dose to which they (and therefore the product to which they were attached) have been exposed. Such dosimeters must be sent in for processing to determine the level and duration of exposure. The measurement consists of analysis on a spectrometer, and then calculations are performed using software to establish the dose. These detectors may only be used once.

Thermoluminescent dosimeters (TLDs) are widely used in industry to measure the amount of energy (dose) per unit mass absorbed by the material when exposed to ionizing radiation. The TLD material absorbs and stores energy when exposed to radiation, and heating the TLD releases the energy as light. Usually personal TLD dosimeter badges are checked on a monthly basis. Feedback is only supplied if a badge is found to indicate the exposure.

Other examples of personal gamma-radiation sensors may include optically-stimulated-luminescence (OSL) dosimeters, film or ring badges, and personal electronic dosimetry cards. Such a nonreal-time approach is acceptable when it is assumed that any accidental exposure would be low level. However, today's industry and safety precautions demand real-time monitoring of any change in the amount of radiation being emitted. Any alteration in the expected level should set off an alarm to prevent any exposure over the safe dose limit.

An illustrative example of a commercially available real-time radiation dosimeter is the GammaMaster watch with a built-in Geiger counter that costs $485 (http://www.gammawatch.com). It displays dose rate (0.01–4,000 μSv/h) as well as cumulative dose (0.001–9,999 μSv). Alarms can be set to indicate when a specified dose rate or cumulative dose is exceeded. In addition to the digital display, the GammaMaster has analog liquid crystal display (LCD), which provide a visual indication of the current dose rate and cumulative dose. The full-scale readings on the analog scales are the alarm settings. In a number of applications, cheaper alternative radiation sensors are needed.

To meet this demand, considerable research into new sensors is underway, including efforts to enhance the sensor performance through both the material properties and manufacturing technologies. The recent availability of various metal oxide materials in high-surface-area nanopowder form, as well as implementation of newly developed nanofabrication techniques, offer tremendous opportunities for various sensors manufacturers. New preparation technologies and optimized deposition process are essential to achieve better control of the materials characteristics and consequently to improve the radiation sensor performance.

Chapter 3 focuses on the theoretical aspects of the optical and the electrical properties of the materials. This knowledge is important in understanding the effects that radiation induces in the properties of the materials. These changes in properties affect the overall device performance. Conduction mechanisms in materials and three types of metal-semiconductor contacts are discussed. The primary interactions between energetic radiation and semiconductors and inorganic insulators results in the loss of energy to their electrons, and this energy is ultimately converted to the form of electron-hole pairs. In this process, known as ionization, the valence band electrons in the solid are excited to the conduction band and are highly mobile, if an electric field is applied [5]. The production and subsequent trapping of the holes in oxide films cause serious alterations in the devices performance.

In organic materials under irradiation, a chain of reactions, in which oxygen and moisture from the environment may be incorporated, starts with rapid electronic phenomena. This is followed by the generation of reactive, short-lived intermediate compounds and chemical products of complex mixture. The main result of ionization is the breaking of chemical bonds and the creation of new ones, which causes change in conductivity and leads to long-lived forms of physical breakdown. Along with radiation damage in crystalline structures and oxide materials, this Chapter covers the radiation effects in polymers.

Numerous efforts are devoted to investigating the influence of radiation on the properties of metal oxides and polymer materials [6–11]. Gamma irradiation ($10^6 - 5 \times 10^7$ rad) degrades the electrical and dielectric properties of thin tantalum pentoxide layers obtained by RF sputtering and thermal oxidation in terms of dielectric constant, oxide charge, and leakage current [6]. The main source of electrically active defects in irradiated films is associated with the oxygen vacancies and the broken Ta–O and Si–O bonds. After the electron and γ-ray radiation of titanium oxide films are prepared by the DC-reactive sputtering method, the number of Ti^{4+} ions decreases in the transition layer, and Ti^{4+} turns into Ti^{3+} in accordance with reaction $2TiO_2 \rightarrow Ti_2O_3 + O$ [7]. The influence of ionizing radiation on NiO and its mixture with other oxides prepared by various techniques has been explored. Irradiation with 1 MSv gamma rays of NiO led to changes in both its surface oxidative abilities and its catalytic activities [8]. Thermal vacuum evaporated thin films of TeO_2 showed an increase in the values of current with the increase in radiation dose [12].

Deep understanding of the physical properties of the materials under the influence of radiation exposure is vital for the effective design of dosimeter devices. Detection of radiation is based on the fact that both the electrical and the optical properties of the materials undergo changes upon the exposure to ionizing radiation. It is believed that radiation causes structural defects (called color centers or oxygen vacancies in oxides) leading to changes in their density upon the exposure to radiation. The influence of radiation depends on both the dose and the parameters of the films, including their thicknesses: the degradation is more severe for the higher doses and the thinner films. Extensive experimental work has been conducted to explore and correlate the effects of the radiation on the electrical, optical, and structural properties of thin/thick films.

Chapter 4 highlights the efforts that were made to investigate the influence of radiation on metal oxides and polymer materials for dosimetry applications. Metal oxides (such as NiO, CeO_2, TeO_2, In_2O_3, SiO, and MnO) and polymers (such as CuPc, NiPc, MnPc, and CoPc) were used as the active constituents in the fabrication of γ-radiation sensors. A detailed investigation into the use of both thin and thick film materials themselves as the primary sensors for gamma radiation is given. Thin/thick film devices were made in the form of resistor- and capacitor-type structures, planar structures with interdigitated electrodes, p-n junctions, and metal-oxide-semiconductor (MOS) devices. Thin and thick film technologies are relatively inexpensive to implement but are not used commercially because they have not been developed to an appropriate level. The major advantages of these techniques are the versatility in materials preparation and the simplicity in film formation, which result in an easy automation of processes, with good miniaturization, repeatability, and reliability, and a low manufacturing cost. Thick and thin film technologies exhibit a number of characteristics that are relevant to the realization of sensors. In particular, they are compact, robust, and cost effective.

Several metal oxides and their mixtures in different proportions in the forms of thin and thick films were used as radiation sensing layers. In the last 20 years, many phthalocyanine derivatives have become important materials for use in diverse fields, such as laser printers and photocopiers, gas sensors, optical logic displays, solar cells, and color filters [13]. These applications were made possible by the unique properties of phthalocyanines, such as their stability at temperatures below 773K, their low sublimation pressure, their chemical stability, and their resemblance in molecular structure to hemoglobin and chlorophyll. In general, phthalocyanines are p-type semiconductors with a large band gap that can be altered by changing the metal constituent. The interaction of ionizing radiation with matter usually causes the formation of short-lived intermediate compounds with a high density of dangling bonds. This radiation-induced damage produces measurable changes in the optical and electrical properties of the materials, which can be used to evaluate the observed dose. Organic compounds are sensitive to radiation by irreversible chemical processes, known as radiolysis, that occur when the C-H and C-C bonds are ruptured to produce new organic compounds [5]. The effects of gamma radiation on the absorbance bands, permittivity, and conduction mechanisms of both thin and thick films of metal-substituted phthalocyanines are discussed in Chapter 4.

In general, thin film devices are more sensitive to lower doses of radiation than their thick film counterparts. It is therefore recommended to use thin film devices for low-dose applications, and thick film devices for higher dose applications, as the latter were found to sustain higher radiation doses. Since it is difficult to develop one sensor that can detect both low and high doses of ionizing radiation, different dose ranges have to be examined and corresponding devices have to be manufactured accordingly. To cover more than one energy or type of radiation, the approach of using devices with a combined structure, such as sensor arrays, can be utilized, where sections of the radiation sensor could differ in material thickness or composition. A few examples of such sensor arrays are given in Chapter 5.

In view of the wide variety of radiation sources employed in industry, medicine, and research, it is desirable that a device used for dosimetry is capable of giving a reasonably accurate estimate of the total dose received from exposure to radiations differing in both energy and type. It is also necessary to have the ability to separate the individual contributions of the different types of radiation that make up the mixture, since the permissible levels for α-, β-, and γ-radiations differ appreciably. Such knowledge is particularly essential if an attempt is made to determine the dose delivered at a depth within a body. It may be of value in cases where an investigation is necessary to determine the cause of an overexposure. Potential applications of the dosimetry of mixed external radiations include cases in which type and level of radiation is unknown, such as building sites, mining, and military and security purposes. Chapter 5 presents a novel concept of radiation nose. In a radiation nose system, an array of sensors is used to monitor various ranges and types of radiation, therefore increasing the working dose range and the sensitivity of the overall system.

Based on a set of output readings from the sensors, a specific PR algorithm should be applied, so that the radiation nose system will accurately determine the radiation dose and type. The overview of PR algorithms, along with feature extraction/selection, classification, and validation, is given in Chapter 5. As an example, a radiation nose system using a Compaq iPAQ designed for resistive sensors is presented. Alternatively, in the case of capacitive sensors, the circuitry and software for a portable real-time gamma radiation dosimetry system is offered. The reader should treat these examples purely as an illustration of tremendous possibilities that engineers, researchers, and designers have in development of novel cost-effective radiation monitoring systems. Depending on the application and the requirements for the dosimetry system in each case, proper radiation-sensitive material and sensor design have to be chosen to meet customer specifications. A wide range of factors has to be accounted for, such as efficiency and resolution requirements and reliability under certain environmental conditions.

The authors hope that this book will provide a general knowledge in radiation dosimetry for a broad audience and offer engineers and researchers working on the development of novel real-time dosimetry systems a place to start. Chapter 6 summarizes the book and speculates on future trends and advancements in radiation sensors, which are toward miniaturization and wireless monitoring.

References

[1] Watson, S. J., et al., *Ionising Radiation Exposure of the UK Population: 2005 Review*, Health Protection Agency Report, Vol. HPA-PRD-001, 2005.

[2] Miller, A., P. Sharpe, and R. Chu, *Dosimetry for Industrial Radiation Processing*, ICRU, 2000.

[3] Lopez Ponte, M. A., et al., "Individual Monitoring for Internal Exposure in Europe and the Integration of Dosimetric Data," *Radiation Protection Dosimetry*, Vol. 112, No. 1, 2004, pp. 69–119.

[4] Audet, S., and J. Steigerwald, "Radiation Sensors," in *Semiconductor Sensors*, S. M. Sze, (ed.), New York: John Wiley & Sons, 1994, pp. 271–329.

[5] Holmes-Siedle, A. G., and L. Adams, *Handbook of Radiation Effects*, New York: Oxford University Press, 1993.

[6] Atanassova, E., et al., "Influence of γ-Radiation on Thin Ta_2O_5-Si Structures," *Microelectronics Journal*, Vol. 32, No. 7, 2001, pp. 553–562.

[7] Zhang, J. D., et al., "Ti Ion Valence Variation Induced by Ionizing Radiation at TiO_2/Si Interface," *Surface and Coatings Technology*, Vol. 158–159, 2002, pp. 238–241.

[8] Mucka, V., J. Podlaha, and R. Silber, "NiO-ThO_2 Mixed Catalysts in Hydrogen Peroxide Decomposition and Influence of Ionizing Radiation," *Radiation Physics and Chemistry*, Vol. 59, No. 5–6, 2000, pp. 467–475.

[9] Arshak, K., et al., "Thin and Thick Films of Metal Oxides and Metal Phthalocyanines as Gamma Radiation Dosimeters," *IEEE Transactions on Nuclear Science*, Vol. 51, No. 5, 2004, pp. 2250–2255.

[10] Arshak, K., and O. Korostynska, "Thin- and Thick-Film Real-Time Gamma Radiation Detectors," *IEEE Sensors Journal*, Vol. 5, No. 4, 2005, pp. 574–580.

[11] Arshak, K., O. Korostynska, and D. Morris, "Mixed and Carbon Filled Oxide Materials as Gamma Radiation Sensors," *Materials Science and Engineering B*, Vol. 118, No. 1–3, 2005, pp. 275–280.

[12] Arshak, K., and O. Korostynska, "Gamma Radiation-Induced Changes in the Electrical and Optical Properties of Tellurium Dioxide Thin Films," *IEEE Sensors*, Vol. 3, No. 6, 2003, pp. 717–721.

[13] Engel, M., "Single-Crystal and Solid-State Molecular Structures of Phthalocyanine Complexes," *Technical Report*, Vol. 1996, pp. 11–54.

2

Radiation Dosimetry: Background and Principles

2.1 Review of Radiation Types

2.1.1 Ionizing Radiation

Radiation that has enough energy to move atoms in a molecule around or cause them to vibrate, but not enough to change them chemically, is referred to as nonionizing radiation. Examples of this kind of radiation are sound waves, visible light, and microwaves. Radiation that falls within the ionizing radiation range has enough energy to actually break the chemical bonds. Nonionizing radiation is used for many common tasks, from heating food to broadcasting, and ranges from extremely low frequency radiation through the audible, microwave, and visible portions of the spectrum into the ultraviolet range. X-ray and gamma ray radiation, which are at the upper end of the electromagnetic spectrum, have very high frequencies of $>10^{19}$ Hz, and they exhibit very short wavelengths of the order of $<10^{-8}$ meters. Radiation in this range has extremely high energy (greater than 10^6 eV).

Ionizing radiation falls into two categories: direct ionizing (charged particles such as electrons and protons) and indirect ionizing (uncharged particles such as photons and neutrons).

Ionization is the process in which a charged portion of a molecule (usually an electron) is given enough energy to break away from the atom. This process results in the formation of two charged particles or ions: the molecule with a net positive charge and the free electron with a negative charge. Any charged particle passing through matter loses energy to the electrons of the atoms it encounters. Energy is transferred between charged particles by electrostatic (coulomb) forces, causing the affected electrons to move into higher orbital energy levels (excitation) or to escape the orbital atomic structure completely (ionization). Each unbound electron may then produce additional excitations or ionizations in other atoms until its energy is expended. Since a charged particle has a very high probability of interacting with each electron that is near its path, the loss of energy is continuous as the particle passes through matter. The rate of energy loss increases as the kinetic energy of the particle decreases until the remaining energy is not sufficient to produce additional excitations or ionizations. The energy expended in raising an electron to an excited state or in releasing it completely from an atom is released as a photon of electromagnetic radiation when the electron returns to its normal energy level. The energy of the emitted photon depends upon the transition experienced by the electron. Minor transitions, such as those from an excited to a normal state within the same general energy level (electron shell), may produce photons of ultraviolet or visible light. Transitions between major energy levels produce photons called *characteristic X-rays*, each having a unique energy representing a difference in the electron binding energies characteristic of the atom.

A charged particle may also lose energy by emission of electromagnetic radiation (photons) during deceleration. The emitted radiation is called bremsstrahlung, a German word meaning *braking radiation*. This form of energy loss occurs predominantly when very energetic electrons interact with a material of high atomic number (e.g., in the target of an X-ray tube). The quantity and energies of the emitted photons increase rapidly with an increased atomic number of the stopping material. The entire kinetic energy of the electron may be converted to a single photon, but usually only a small fraction of the energy is transferred to a photon. When beta particles from ^{32}P interact with lead, up to 7% of the total beta energy emitted is converted to bremsstrahlung, with average photon energy of 35 keV and a maximum energy of 1.7 MeV.

2.1.1.1 Alpha Particles

Discovered by Ernest Rutherford in 1899, alpha particles are a type of directly ionizing radiation ejected by the nuclei of some unstable atoms. They are large subatomic fragments consisting of two protons and two neutrons, identical to a helium nucleus. An alpha particle is a relatively heavy, high-energy particle, with a positive charge of +2 from its two protons. The scheme of alpha decay is shown in Figure 2.1. Alpha particles are emitted to restore balance when the ratio of neutrons to protons in the nucleus is too high. The nucleus is initially in an unstable energy state. An internal change takes place in the unstable nucleus, and an alpha particle is ejected, leaving a decay product. The atom has then lost two protons along with two neutrons. Since the number of protons in the nucleus of an atom determines the element, the loss of an alpha particle actually changes the atom to a different element. Alpha-emitting atoms tend to have

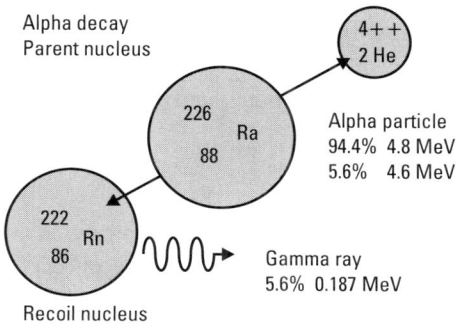

Figure 2.1 Alpha decay.

high atomic numbers. Alpha rays from radionuclides with energies of a few MeV travel only a few inches in air and are easily shielded by skin, clothing, or a piece of paper. External alpha irradiation of the human body is therefore of little concern. However, ingested radionuclides can lead to direct alpha exposure of critical cells in organs, such as the lung and bone. Alpha rays are dangerous because they deposit their energy in a small volume, thus causing great damage to a few cells.

2.1.1.2 Beta Decay

Beta particles are electrons emitted during nuclear decay process, which is schematically written as [1]:

$$_Z^A X \rightarrow _{Z+1}^A Y + \beta^- + \bar{\nu} \qquad (2.1)$$

where X and Y are the initial and final nuclear species, and $\bar{\nu}$ is the antineutrino. Beta particles have the same mass as an electron but may be either negatively or positively charged. With their small size and charge, they penetrate matter more easily than alpha particles but are more easily deflected [2].

2.1.1.3 Gamma Radiation and X-Rays

Gamma ray emission often accompanies the emission of a beta particle. When the beta particle ejection does not liberate the nucleus of the extra energy, the nucleus releases the remaining excess energy in the form of a gamma photon. A gamma photon is a packet of electromagnetic energy having no mass and no electrical charge. Henri Becquerel is also credited with discovering of gamma radiation in 1896. Gamma photons are the most energetic photons in the electromagnetic spectrum and can range from a few keV up to around 10 MeV. While gamma rays and X-rays pose the same hazard, they differ in their origin. Gamma rays originate in the nucleus. X-rays originate in the electron fields surrounding the nucleus. Gamma radiation emission occurs when the nucleus of a radioactive atom has too much energy and often follows the emission of a beta particle. Atoms emit X-rays when electrons fall from a higher energy shell to a lower energy shell.

2.1.1.4 Neutrons

A neutron has the same mass as a proton but has no charge and is difficult to stop. The capture of a neutron results in the emission of a gamma ray. Neutrons are classified according to their energy: thermal (< 1 eV); intermediate; and fast (> 100 keV). Water is regarded as an effective shield for neutrons. Table 2.1 compares some key characteristics of the ionizing radiations with $E = 1$ MeV.

2.1.2 Stopping Power and Linear Energy Transfer

2.1.2.1 Stopping Power

For electrons of energy less than 5 MeV, almost all energy loss on passing through the material is by interaction mechanisms that result in an ionization of the material (i.e., the creation of electron-hole pairs with little momentum transfer to the atoms). The rate of loss of energy E with distance traversed is known as the stopping power of the material and is given by [2]:

$$-\frac{dE}{dx} = \frac{2e^4 z^2 N_A Z}{mv^2 A} B \tag{2.2}$$

Table 2.1
Comparison of Ionizing Radiations with $E = 1$ MeV

	Alpha (α)	Proton (p)	Beta (β) or Electron (e)	Photon (γ- or X-Ray)	Neutron (n)
Symbol	$^4_2\alpha$ or He^{2+}	^1_1p or H^{1+}	$^0_{-1}e$ or β	$^0_0\gamma$	^1_0n
Charge	+2	+1	−1	Neutral	Neutral
Ionization	Direct	Direct	Direct	Indirect	Indirect
Mass (amu)	4.00277	1.007276	0.000548	—	1.008665
Velocity (cm/s)	6.944×10^8	1.38×10^9	2.82×10^{10}	2.998×10^{10}	1.38×10^9
Speed of light	2.3%	4.6%	94.1%	100%	4.6%
Range in air (cm)	0.56	1.81	319	82,000	39,250

where N_A/A here is the number of atoms of atomic number Z per unit volume ($N_A = 6 \times 10^{23}$ atoms per mole); A is the atomic mass; z and v are the charge number (1 for an electron) and velocity of the incident particle, respectively; and x is the path length or distance measured along the track of an electron. B is known as the stopping power of the material and varies slowly with particle energy as follows [1]:

$$B = Z\left[\ln\frac{2mv^2}{I} - \ln\left(1 - \frac{v^2}{c^2}\right) - \frac{v^2}{c^2}\right] \quad (2.3)$$

The parameter I represents the average excitation and ionization potential of the absorber and is normally treated as an experimentally determined parameter for each element. The minimum stopping power for electrons occurs at energies in the range of 1–2 MeV. Note that stopping power is often quoted in units of energy lost per unit mass thickness, measured along the particle path in MeV cm^2 g^{-1}. In addition to the energy loss by collision, there is a further contribution to stopping power due to radiation loss (bremsstrahlung generation). At energies below 1 MeV, this is extremely small when compared with collision loss as a mechanism for stopping electrons. It is a rising function of energy, but it does not dominate over collision loss until the energies exceed well over 10 MeV [2].

2.1.2.2 Particle Range

The mean range of the alpha particle in the absorber material is defined as the absorber thickness, which reduced the alpha particle count to exactly one half of its value in the absence of the absorber [1]. The range of charged particles of a given energy is thus a fairly unique quantity in a specific absorber material. Knowing the particle range is extremely important when designing radiation detectors. Any detector that has to measure the full incident energy of a charged particle must have an active thickness, which is greater than the range of that particle in the detector material.

International Commission on Radiation Units and Measurements (ICRU) Report 49 encompasses the tabulation of stopping powers and ranges for 73 materials and covers the energy ranges 1 keV to 10,000 MeV for protons, and up to 1,000 MeV for alpha particles [3]. The materials covered are those of interest in radiological physics and biomedical dosimetry. They range through such elements as He, Al, Fe, Au, and U and such compounds and mixtures as acetylene, muscle (skeletal), polyethylene, and tissue-equivalent gas. The stopping power tables span more than 140 pages in Report 49. Such matters as calculation of electronic stopping powers of protons and alpha particles at high energies according to Bethe's theory [4] with various corrections are also covered in some detail. The use of experimental information for electronic stopping powers at low energies is also treated. Sections of Report 49 cover calculations of nuclear stopping power and comparisons of tabulated values with experimental results. Finally, the report also provides concise descriptions of methods used for stopping power measurements. Figures 2.2 through 2.5 illustrate stopping powers for

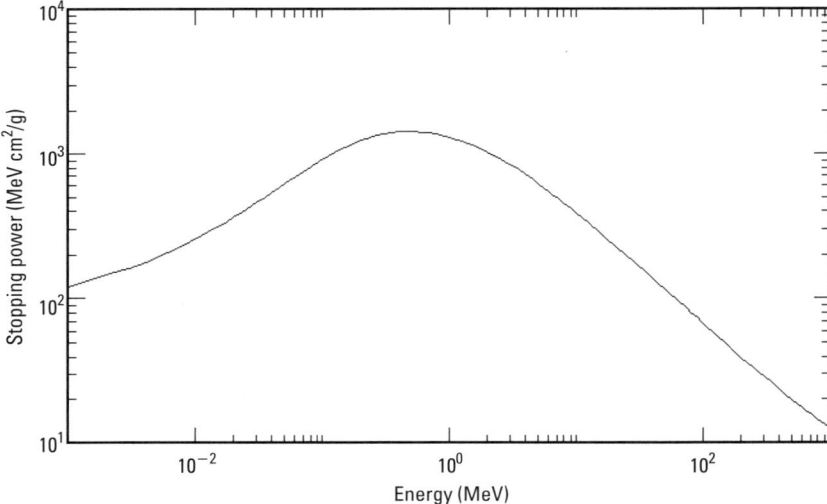

Figure 2.2 Stopping power for alpha particles in silicon.

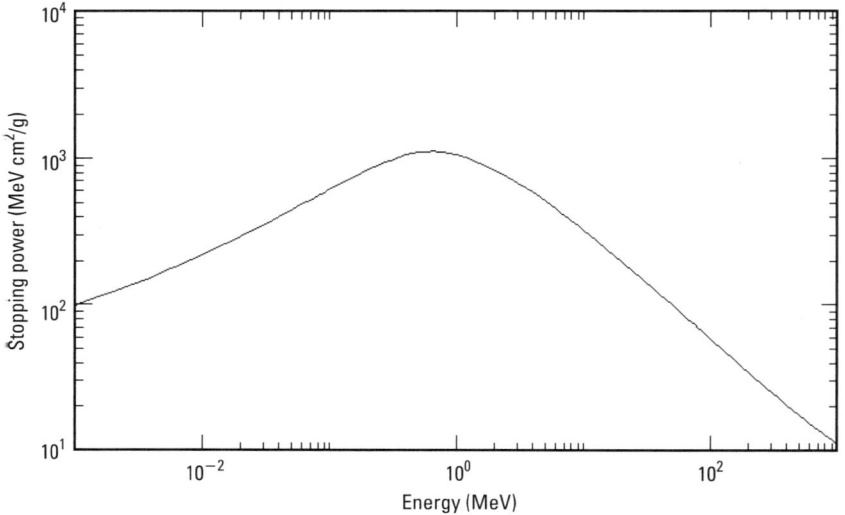

Figure 2.3 Stopping power for alpha particles in titanium.

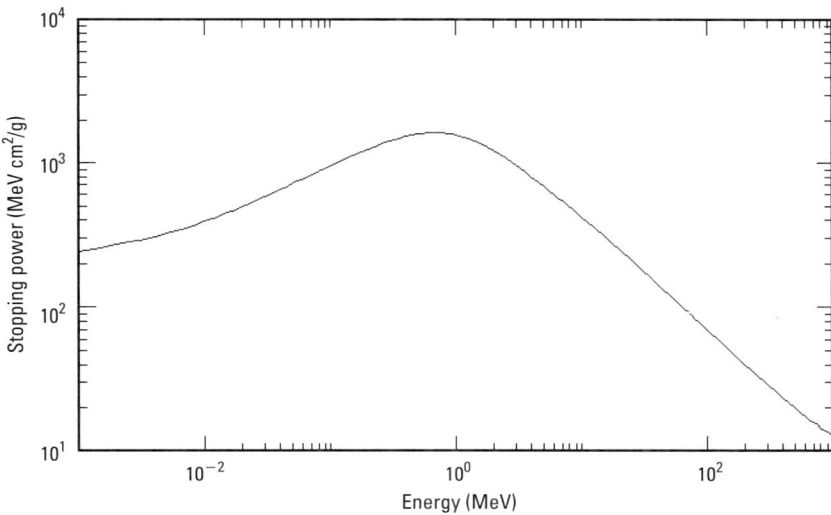

Figure 2.4 Stopping power for alpha particles in lithium fluoride.

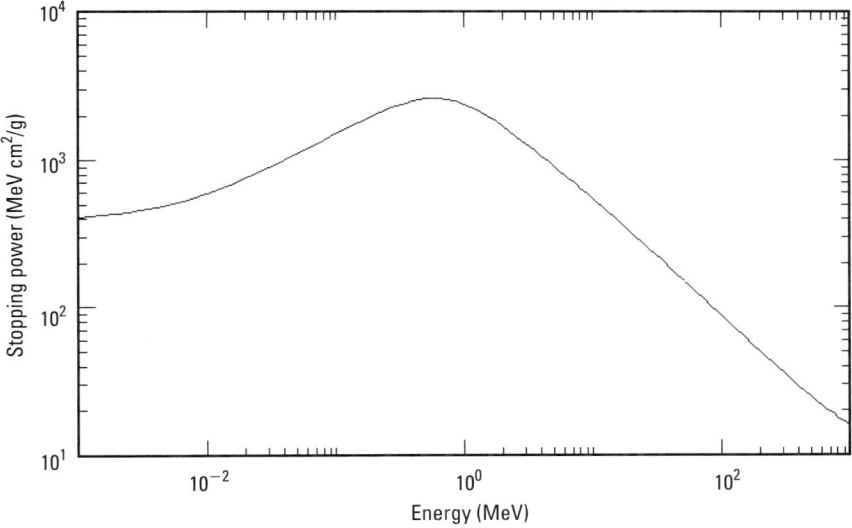

Figure 2.5 Stopping power for alpha particles in A-150 tissue-equivalent plastic.

alpha particles in silicon, titanium, lithium fluoride and A-150 tissue-equivalent plastic.

Printed tables remain indispensable as a means of quickly looking up information, but there is an increasing need for computer-readable databases that can quickly generate the desired data. To meet such a need, the National

Institute of Standards and Technology (NIST) Physics Laboratory (http://physics.nist.gov/PhysRefData/Star/Text/contents.html) developed Web databases ESTAR, PSTAR, and ASTAR. With a default option, PSTAR and ASTAR generate the stopping powers and ranges for protons and helium ions tabulated in ICRU Report 49 [3] for 73 materials at a standard grid of 133 kinetic energies between 1 keV and 10 GeV for protons, and 122 kinetic energies between 1 keV and 1 GeV for helium ions. These databases can also calculate similar results at any other energy grid between these limits.

ICRU Report 37 represents the first result of a series of Commission efforts concerned with stopping power for electrons and positrons [5]. In treating electrons, the report focuses on radiative stopping power due to the emission of bremsstrahlung and gives information on electron collision stopping power at energies below 10 keV. Stopping power tables are presented for a large number of elements and compounds covering the energy range from 10 keV to 1,000 keV. The tables also include the range and radiation yield. Throughout the report, the requirements for up-to-date stopping power information in biomedical dosimetry are emphasized. Major sections of the report discuss formulas for collision stopping power, methods for estimating mean excitation energies, selection of mean excitation energies for elements and compounds, density effect, restricted collision stopping power, electron collision stopping powers at low energies, radiative stopping power, ranges, and radiation yields [5].

With a default option, ESTAR generates stopping powers and ranges for electrons, which are the same as those tabulated in ICRU Report 37 for 72 materials at a standard grid of 81 kinetic energies between 10 keV and 1,000 MeV (http://physics.nist.gov/PhysRefData/Star/Text/ESTAR.html). ESTAR can also calculate similar tables for any other element, compound, or mixture. Furthermore, it can calculate stopping powers at any set of kinetic energies between 1 keV and 10 GeV. Figures 2.6 through 2.9 illustrate stopping powers for electrons in silicon, titanium, lithium fluoride and A-150 tissue-equivalent plastic.

For comparison, Figure 2.10 illustrates linear stopping power of different particles in silicon.

2.1.2.3 Linear Energy Transfer

Damage to biological tissue by ionizing radiation is caused by energy absorption and the resulting ionization and excitation processes in the biological microstructures. The energy transfer to atom and atomic nuclei can result in biochemical reactions with the molecules as well as in biological changes to cells or cell components. In addition to direct energy transfer by interaction, energy can be transferred indirectly to a molecule by radicals or their reaction products, such as H_2O_2. Depending on the irradiation conditions, damage to the organism can result from the energy transfer and may extend to single organs, organ systems, or the whole body. The absorption of equal amounts of energy per unit mass

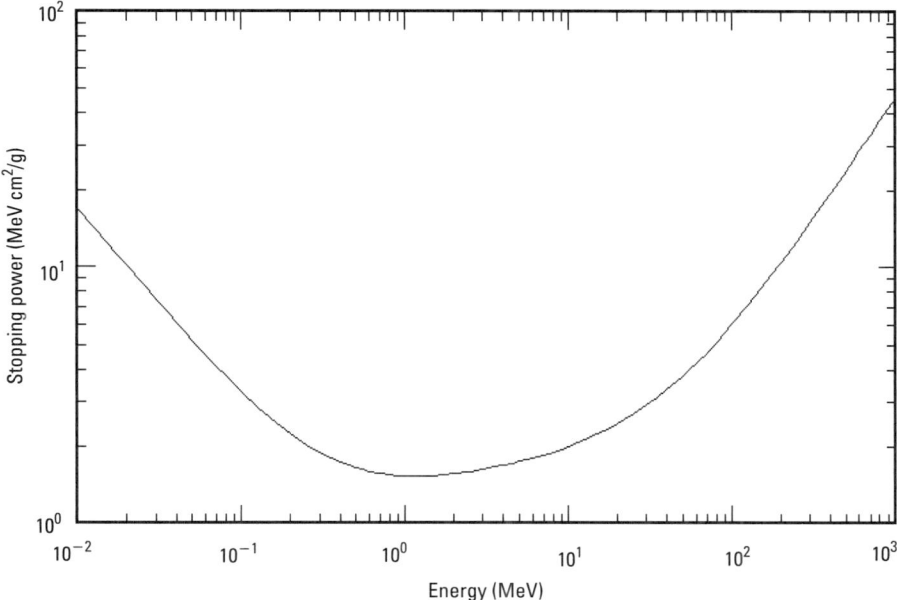

Figure 2.6 Stopping power for electrons in silicon.

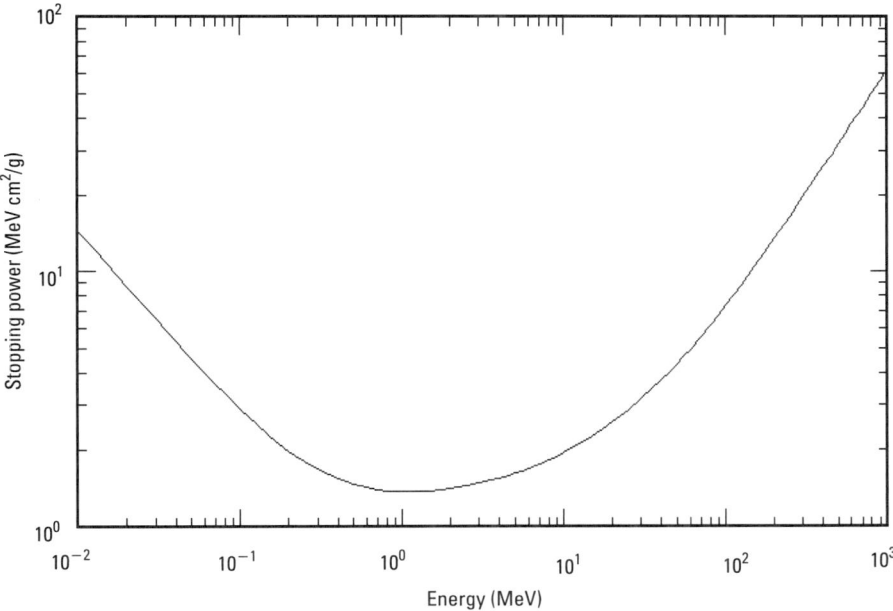

Figure 2.7 Stopping power for electrons in titanium.

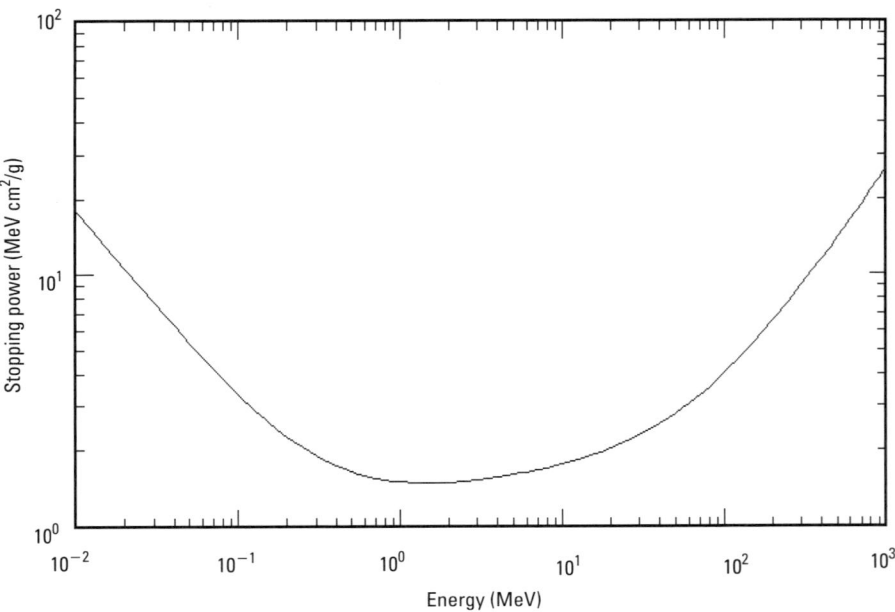

Figure 2.8 Stopping power for electrons in lithium fluoride.

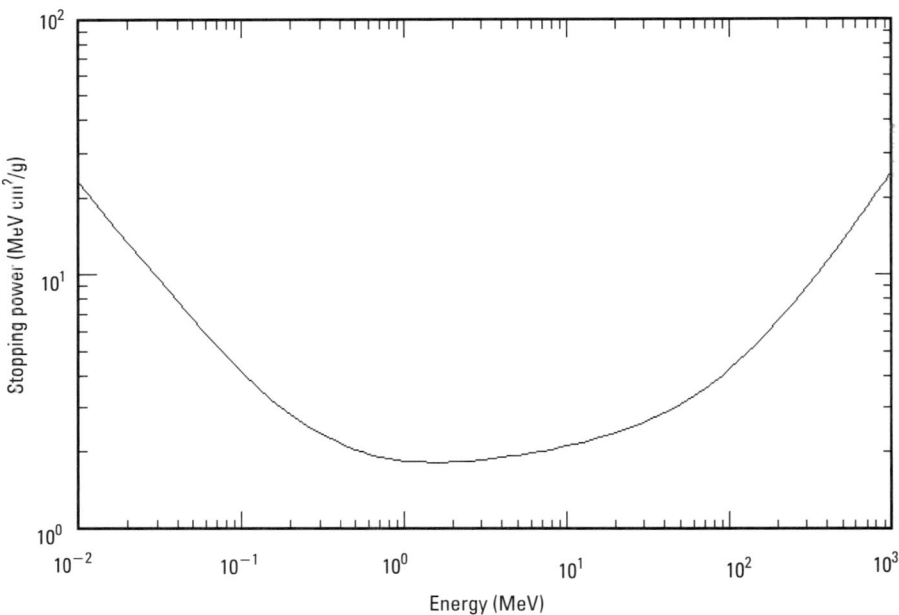

Figure 2.9 Stopping power for electrons in A-150 tissue-equivalent plastic.

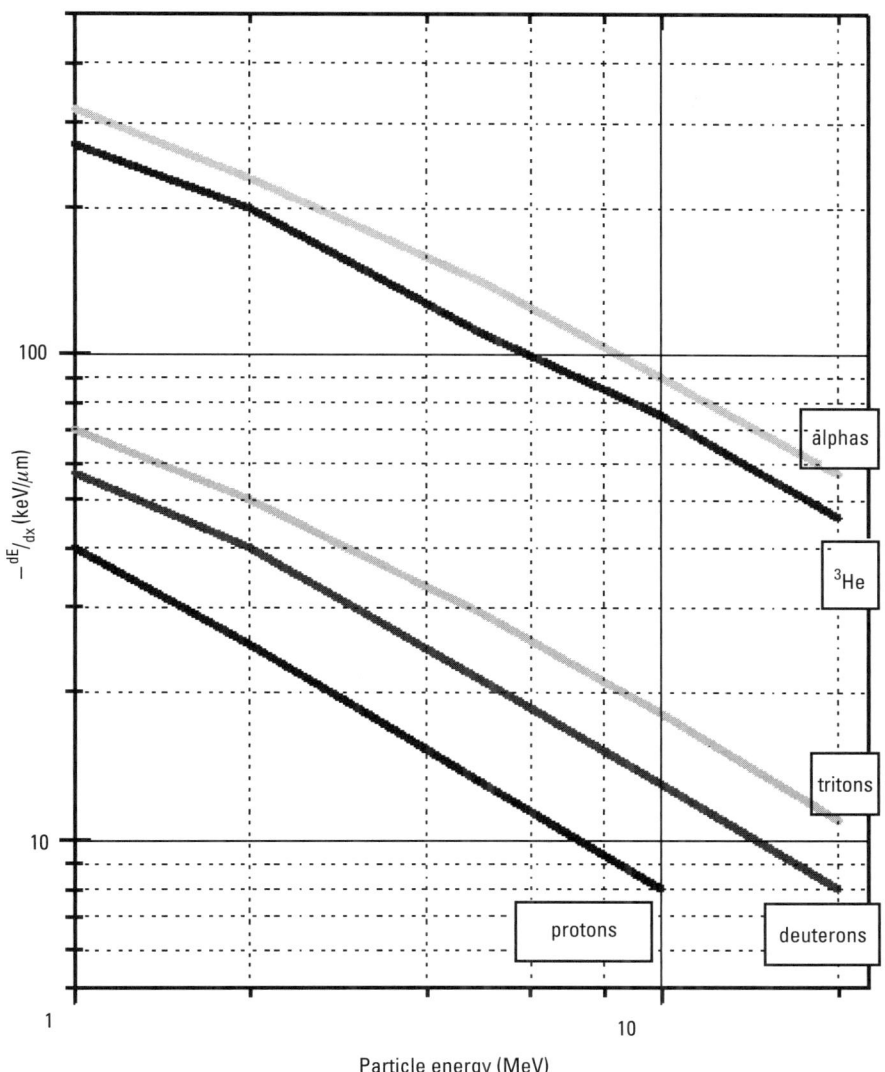

Figure 2.10 Linear stopping power as a function of particle energy in silicon.

under different irradiation conditions would not have the same biological effect. The severity and permanence of these changes are directly related to the local rate of energy deposition along the particle track, known as linear energy transfer (LET) [1]:

$$L = \frac{dE_L}{dl} \qquad (2.4)$$

where dE_L is the average energy locally imparted to the medium. Radiations with large linear energy transfer result in a greater biological damage than those with lower linear transfer.

ICRU Report 16 includes major sections on the interaction of radiation with matter, definition and concepts of LET, calculation of distributions of absorbed dose in LET, applications of LET calculations, LET in radiation protection, limitations of the LET concept, and other methods of specifying radiation quality [6]. The report also includes appendixes covering formulas for mass collision stopping power, theoretical and experimental values for range, total average rate of energy loss per unit path length and LET, measurement of LET distributions, the mean excitation energy, and the application of LET in radiation biology and chemical dosimetry.

Biological tissue is often approximated by so-called standard tissue with the following composition: 10.1% of H, 11.1% of C, 2.6% of N, and 76.2% of O [7]. The corresponding LET values for electrons, protons, and alpha particles at different energy levels are given in Table 2.2.

2.1.2.4 Dose Equivalent

The concept of dose equivalent is used to quantify the probable biological effect of a given radiation exposure. A unit of dose equivalent is defined as the amount of any type of radiation that, when absorbed in a biological system, results in the same biological effect as one unit of absorbed dose delivered in the form of low linear energy transfer radiation. The dose equivalent H is the product of the

Table 2.2
LET in Standard Tissue at Various Particle Energies

Energy (MeV)	LET (keV μm^{-1})		
	Electrons	Protons	Alpha Particles
0.0001	55.2		
0.001	7.48		
0.01	0.98		
0.1	0.152		
1	0.060	13.9	182
2		7.6	101
4		4.12	55.4
6		2.87	38.9
8		2.23	30.1
10		1.83	24.7

absorbed dose D and the quality factor Q, which characterizes the specific radiation. The Q value increases with increasing linear energy transfer. For the fast electron radiations, LET is sufficiently low; therefore for beta and gamma radiations, Q is 1. Alpha radiation is far more damaging per unit of energy deposited in living tissue. The quality factor for alpha radiation was taken as 20 [1]. The quality factor used for neutrons is 10.

The units of dose equivalent H depend on the units of the absorbed dose D. The absorbed dose is defined in terms of energy that is absorbed from any type of radiation per unit mass of the absorber. D was historically expressed in rads, whereas H was defined in rems. Under SI (International System of Units) convention, D is expressed in gray (Gy) and a corresponding unit of dose equivalent is called the sievert (Sv).

$$1 \text{ Gy} = 1 \text{ Jkg}^{-1} = 100 \text{ rad}$$
$$1 \text{ rad} = 100 \text{ erg g}^{-1} = 6.25 \times 10^{13} \text{ eVg}^{-1} = 10^{-2} \text{ Gy}$$

For measurement purposes, the operational quantities of ambient dose equivalent, directional dose equivalent, and personal dose equivalent are defined. Where doses are estimated from area-monitoring results, the relevant operational quantities are ambient dose equivalent and directional dose equivalent.

- *Ambient dose equivalent $H^*(d)$*. The ambient dose equivalent $H^*(d)$, at a point, is the dose equivalent that would be produced by the corresponding expanded and aligned field, in the ICRU sphere at a depth d in millimeters on the radius opposing the direction of the aligned field. For measurement of strongly penetrating radiations, the reference depth used is 10 mm and the quantity is denoted $H^*(10)$.

- *Directional dose equivalent $H'(d,\Omega)$*. The directional dose equivalent $H'(d,\Omega)$, at a point, is the dose equivalent that would be produced by the corresponding expanded field in the ICRU sphere at a depth d on a radius in a specified direction. Directional dose equivalent is of particular use in the assessment of dose to the skin or eye lens.

- *Personal dose equivalent $H_p(d)$*. The personal dose equivalent $H_p(d)$ is the dose equivalent in soft tissue, at an appropriate depth d below a specified point on the body. $H_p(d)$ can be measured with a detector, which is worn at the surface of the body and covered with an appropriate thickness of tissue-equivalent material. For weakly penetrating and strongly penetrating radiation, the recommended depths are 0.07 mm and 10 mm, respectively, although other depths may be appropriate in particular cases (e.g., 3 mm for the lens of the eye). In order to simplify the notation, d is assumed to be expressed in millimeters, and hence the personal dose equivalents at the two recommended depths are denoted by $H_p(0.07)$ and $H_p(10)$. $H_p(10)$, which is the personal dose equivalent at a 10-mm depth, is used to provide an estimate of the effective dose that avoids both underestimation and excessive overestimation. The sensitive cells of the skin are considered to be between 0.05 and 0.1 mm below the skin surface; therefore, $H_p(0.07)$ is used to estimate the equivalent dose to skin.

2.1.3 Radiation Units

A detailed description of all the units that are used in various areas related to radiation and their relations can be found in NCRP Report 82 [8]. Table 2.3 summarizes the units that are used most frequently in measuring ionizing radiation and radiation dose.

Table 2.3
Units Used in Measuring Ionizing Radiation

Quantity	Definition	Units
Absorbed dose	The amount of energy deposited per unit mass.	SI unit: gray (Gy) 1 Gy = 1 J/kg Historic unit: rad 1 rad = 100 erg/g 100 rad = 1 Gy
Equivalent dose	The dose equivalent H is the product of the absorbed dose D and the quality factor Q, which characterizes the specific radiation.	SI unit: sievert (Sv) Historic unit: rem (roentgen equivalent man) 1 Sv = 100 rem rem = rad×Q
Radioactivity	Curie (Ci) is the traditional unit of radioactivity and equals the radioactivity of 1 gram of pure ^{226}Ra. Becquerels (Bq) is the SI unit of radioactivity equal to 1 disintegration per second.	1 Ci = 37 billion Bq 1 nCi = 37 Bq 1 Bq = 27 pCi
Gamma ray exposure	The SI unit of exposure is coulomb per kilogram (C/kg). The historical unit of gamma ray exposure has been the roentgen (R), defined as the exposure, which results in the generation of 1 electrostatic unit of charge per 0.001293g (1 cm^3) of air.	1R = 2.58 × 10^{-4} C/kg 1R = 86.9 erg g^{-1} (air) 1R of 1 MeV photons = 1.95 × 10^9 photons cm^{-2}

2.2 Biological Effects of Radiation

Current knowledge about the biological effects of radiation is based on many sources, including extensive research with animals. Much of the most pertinent data for radiation protection, however, has been obtained from epidemiological studies of human populations exposed inadvertently. Early workers with X-rays and radium, both for medical and commercial applications, were exposed to radiation doses much larger than are permitted today. Patients were also treated with radiation for a variety of illnesses before the possible delayed effects were fully appreciated. Uranium miners received excessive exposures to airborne radioactivity before the introduction of protective regulations and control methods. The other major groups that were exposed significantly and have been studied intensively are the Japanese population that survived the atomic bombings of Hiroshima and Nagasaki, as well as Chernobyl survivors. All these populations were large enough, and received large radiation doses, to provide statistically significant data on the incidence of radiation-induced effects. The types and durations of radiation exposures to these groups were also sufficiently varied to provide a database that includes external exposures to X-rays, gamma rays, and neutrons, and internal exposures from ingested and inhaled alpha- and beta-particle emitters over intervals ranging from days to decades. Since many of the victims of these exposures are still alive, investigations that began 30 to 40 years ago are continuing today.

The organizations that provide the primary scientific evaluations of radiation doses and risks are the National Academy of Sciences Committee on the Biological Effects of Ionizing Radiation (BEIR Committee), the United Nations Scientific Committee on the Effects of Atomic Radiation (UNSCEAR), the ICRP, and the NCRP in the United States. These bodies recommend norms for permissible doses of radiation and determine safe dose limits for radiation workers and for the general public. According to them, annual equivalent dose limits for skin and lens of the eye are 500 mSv and 150 mSv, respectively. More detailed data can be found for example in NCRP Report 136 [9]. ICRP recommendations are given in ICRP publication 60 [10].

2.2.1 Types of Radiation Effects

Ionizing radiation damages cells, primarily through the damage to deoxyribonucleic acid (DNA). Subtle damage to DNA produces mutations, which can lead to cancer in those exposed or genetic effects in offspring if the germ cells in the parents are exposed. Gross damage to DNA can produce cell death. At low doses, cancer risk is predominant. At high doses, cell killing predominates. Cancer risk is considered by ICRP/NCRP to be finite even at the lowest doses. Cancer and genetic effects are referred to as stochastic effects. The dose response relationship for stochastic effects is considered to be linear down to zero dose; it is called linear nonthreshold (LNT) hypothesis. The ICRP sets dose limits for stochastic effects (20 mSv/y for workers; 1 mSv/y for members of the public). At high doses, multiple cell killing can compromise the function of multicellular organs and lead to deterioration of organ function, which may sometimes be life threatening. Such effects are referred to as deterministic, since they exhibit threshold doses below which deleterious effects are not observed. Above this threshold, the probability and the severity of the effect increases with dose. Examples of deterministic effects include death due to large whole body dose, cataract induction in the eye lens (not life threatening), erythema of the skin, skin ulceration, and lung fibrosis. ICRP sets special dose limits for the skin (500 mSv/y for workers; 50 mSv/y for the general public) and eyes (150 mSv/y for workers; 15 mSv/y for the general public) so that deterministic effects are prevented. Special dose limits are not necessary for other organs since the limit for stochastic effects of 20 mSv/y (for workers) automatically prevents deterministic effects on other organs.

Next to cancer, the most common image of the effects of radiation is the creation of mutations. Massive damage leads to cell death. Only minor damage will alter a single gene or set of genes so that the resulting organism reveals the effects of the damage. In the case of higher organisms (mammals), the cellular machinery is highly sophisticated so that much routine damage is repaired. Gonadal tissue in males repairs radiation damage very quickly (days if not hours). In mammalian cells, a membrane protects the nucleus of the cell at all times except during cell division. This membrane effectively reduces the effect of radiation damage in the cellular fluid, as hydrogen peroxide does not effectively penetrate. However, when cells undergo division, this membrane dissolves and the nucleus is exposed to the toxins. Cells that undergo rapid division are thus more sensitive to damage by radiation. In order of their rate of division, these include gonads (males), uterine (females), stomach and intestine lining, and bone marrow (blood production). Tissues such as muscles and nerves undergo division slowly, so they are not as easily damaged. Lung cells are at risk of a number of toxins since they are living cells exposed directly to environmental materials. A growing fetus is an example of an organism in which cells are rapidly dividing. This is why pregnant females must be particularly careful with their exposure to any form of radiation. The fetus is particularly sensitive to radiation, and terrible birth defects or death can be caused by high-dose exposure. Potential biological effects depend on how much and how fast a radiation dose is received. Radiation doses can be grouped into two categories: acute and chronic dose.

Acute doses can cause a pattern of clearly identifiable symptoms (syndromes). These conditions are referred to in general as acute radiation syndrome. Radiation sickness symptoms are apparent following acute doses ≥ 1 Gy. Acute whole body doses of ≥ 4.5 Gy may result in a statistical expectation that 50% of the population exposed will die within 60 days without medical attention. As with most illnesses, the specific symptoms, the therapy that a doctor might prescribe, and the prospects for recovery vary from one person to another and are generally dependent on the age and general health of the individual.

Blood-forming organ (bone marrow) syndrome (>1 Gy) is characterized by damage to cells that divide at the most rapid pace (such as bone marrow, the spleen, and lymphatic tissue). Symptoms include internal bleeding, fatigue, bacterial infections, and fever.

Gastrointestinal tract syndrome (>10 Gy) is characterized by damage to cells that divide less rapidly (such as the linings of the stomach and intestines). Symptoms include nausea, vomiting, diarrhea, dehydration, electrolytic imbalance, loss of digestion ability, bleeding ulcers, and the symptoms of blood-forming organ syndrome.

Central nervous system syndrome (>50 Gy) is characterized by damage to cells that do not reproduce, such as nerve cells. Symptoms include loss of coordination, confusion, coma, convulsions, shock, and the symptoms of the blood forming organ and gastrointestinal tract syndromes. Scientists indicate that death under these conditions is not caused by actual radiation damage to the nervous system, but rather from complications caused by internal bleeding, as well as fluid and pressure buildup on the brain.

Other effects from an acute dose include:

- 2 to 3 Gy to the skin can result in the reddening of the skin (erythema), similar to a mild sunburn, and may result in hair loss due to damage to hair follicles.
- 6 Gy to the ovaries or testicles can result in permanent sterilization.
- 0.5 Gy to the thyroid gland can result in benign (noncancerous) tumors.

A chronic dose refers to a dose received over a long period of time. The body is better equipped to tolerate a chronic dose than an acute dose. The body has time to repair damage caused by chronic exposures because a smaller percentage of the cells need repair at any given time. The body also has time to replace dead or nonfunctioning cells with new, healthy cells. This is the type of dose received as occupational exposure.

Radiation could be taken into the body due to radionuclides through inhalation, ingestion, or other means. It is generally much harder to estimate doses from substances inside the body. Conventional dosimeters such as film badges and TLD-based devices measure external doses. Internal doses must be calculated using metabolic models. The intake (via inhalation, ingestion, or skin uptake) can be evaluated via air sampling. The transport around the body can be modeled, and absorbed and equivalent doses can be calculated to exposed organs. The effective dose can then be evaluated. The excretion of radioactive material in urine or feces is also used as an indicator of uptake. The size of an internal dose will depend on the chemical and physical form of the material, its pathways and distribution in the body, and the rate of its elimination from the body (called *biological half-life*). Since metabolic factors vary considerably from one person to the next, the internal dose that any individual gets from a particular radionuclide may differ from the dose calculated using its average biological half-life [11]. The effect of dose rate is taken into account in the derivation of ICRP dose limits.

2.2.2 Relative Biological Effectiveness

As mentioned previously, the biological effect of radiation depends on the dose and on the radiation quality. The investigation of radiation effects by a certain type x of radiation takes place mostly by comparison with the same effects produced by the standard radiation (usually γ- or X-rays). The relative biological effectiveness (RBE) serves as a comparison quantity and is defined by the relationship (2.5), which is based on the fact that the doses D_γ and D_x cause the same effect ($E_x = E_\gamma$) [7]:

$$f_{RBE} = \frac{D_\gamma}{D_x}\bigg|_{E_x=E_\gamma} \quad (2.5)$$

From the experimental investigations, it follows that the RBE factor increases with increasing LET, reaches maximum at $L \approx L_0 \approx 110 \text{ keV}\mu\text{m}^{-1}$ and then decreases again. For the microscopic RBE factor f^*_{RBE}, which describes the damage to a biological structural unit, the following approximation can be used:

$$f^*_{RBE} \sim \{1 - \exp[-(L/L_0)2]\}/L \quad (2.6)$$

As one may see, $f^*_{RBE} \sim L$ for $L \leq L_0$. It has to be considered when describing the damage to major regions of tissue or the body that a LET spectrum $D_L(L)$ equation here also develops for the incidence of monoenergetic radiation due to the interaction processes. To describe the damage in a certain tissue volume or organ, the dose fraction per LET interval can be derived as a function of LET and macroscopic RBE factor [7]:

$$f_{RBE} = \frac{\int D_L(L) f^*_{RBE}(L) dL}{\int D_L(L) dL} \quad (2.7)$$

RBE factors depend on the dose, dose rate, and the effect considered. They are suitable for characterizing radiation effect in radiation biology, radiation therapy, or in the case of radiation accidents. They cannot be applied to characterize the radiation effect at small doses and dose rates, which are typical of radiation exposure of the general population. In this case, the macroscopic quality factor f_Q or Q has to be used.

Since both the macroscopic quality factor and the absorbed dose depend on the LET spectrum in tissue, and this LET spectrum is a function of position in the body, the dose equivalent also depends on the reference point in the body. Weighting factors w_T for several organs and tissues T have been fixed for characterizing the radiation exposure of man in ICRP publication 60 [10]. These weighting factors result from the proportionate radiation risk R_T for the particular organ. Numerical data for R_T and the weighting factors are given in Table 2.4.

The effective dose equivalent H_E can be defined using the weighting factors w_T and weighted dose equivalents in the organs or tissue H_T as:

$$H_E = \sum_T w_T H_T \tag{2.8}$$

Table 2.4
Risks R_T and Tissue Weighting Factors w_T for Determining the Effective Dose According to ICRP 60

Organ or Tissue	$R_T (10^{-4} \, Sv^{-1})$	w_T
Gonads	60	0.20
Red bone marrow	40	0.12
Colon	68	0.12
Lung	68	0.12
Stomach	88	0.12
Bladder	24	0.05
Breast	16	0.05
Liver	12	0.05
Oesophagus	24	0.05
Thyroid	6	0.05
Skin	2	0.01
Bone surfaces	4	0.01
Remainder	40	0.05
Total	$4.6 \times 10^{-2} \, Sv^{-1}$	1.00

Source: [10].

The weighting factors represent mean values for all age groups and both sexes. A measure of radiation exposure has therefore been defined with the effective dose equivalent, which does not serve to strictly determine the actual radiation exposure of an individual but represents the mean radiation exposure of a member of a mixed population.

In its 1990 recommendations, the ICRP introduced a modified concept [10]. For radiological protection purposes, the absorbed dose is averaged over an organ or tissue, T, and this absorbed average dose is weighted for the radiation quality in terms of the *radiation weighting factor* w_R for the type and energy of radiation incident on the body [12]. The resulting weighted dose is designated as the organ- or tissue-equivalent dose H_T according to:

$$H_T = \sum_R w_R D_{T,R} \qquad (2.9)$$

$D_{T,R}$ represents the applied absorbed dose in the organ T by R type of radiation. The values of radiation weighting factors w_R depend only on the type and energy of radiation and are based on present knowledge of RBE factors. Since 1990, there have been substantial developments in biological and dosimetric knowledge that justify a reappraisal of w_R values, which can be found in Table 2.5 [10].

The sum of the organ equivalent doses H_T weighted by the organ weighting factors w_R is termed as the *effective dose E*:

Table 2.5
Radiation Weighting Factors w_R According to ICRP 60

Type of Radiation and Energy Range		w_R
Photons		1
Electrons and muons		1
Neutrons	$E < 10$ keV	5
	10 keV $\leq E \leq 100$ keV	10
	100 keV $< E \leq 2$ MeV	20
	2 MeV $< E \leq 20$ MeV	10
	$E > 20$ MeV	5
Protons (except recoil protons) $E > 2$ MeV		5
Alpha particles, fission fragments, heavy nuclei		20

Source: [10].

$$E = \sum_T w_T \sum_R w_R D_{T,R} \qquad (2.10)$$

ICRP 60 discusses in detail the values of RBE with regard to stochastic effects, which are central to the selection of w_R and Q [12]. Those factors and the dose-equivalent quantities are restricted to the dose range of interest to radiation protection (i.e., to the general magnitude of the dose limits). In special circumstances where one deals with higher doses that can cause deterministic effects, the relevant RBE values are applied to obtain a weighted dose. ICRP has recently reviewed the choice of w_R values and may change them slightly in forthcoming recommendations.

2.3 Basic Principles of Radiation Protection

The basic task of radiation protection consists of avoiding undue exposure of man and the environment to ionizing radiation [7]. Taking into consideration that the probability of stochastic and deterministic damage is dose-dependent, there are two general criteria:

1. Radiation protection measures have to guarantee that deterministic radiation damage is avoided. Since deterministic damage appears above a threshold dose, this dose must not be exceeded.
2. The probability of stochastic radiation damage, which is not zero even at low doses according to the dose-effect relationships currently taken as a basis, must not exceed a justifiable size.

The ICRP initially published its general recommendation for radiological protection in 1928. Further recommendations adapted to the latest information followed in 1959 and 1966. Since 1977, when the ICRP published its fundamental recommendations as ICRP Publication 26, the ICRP has checked these recommendations annually and published supplementary comments from time to time in the annals of the ICRP. Publication 60 is the last statement of general recommendations and it is in the process of being updated at the moment. The focus is on the research needs for setting radiation protection guidelines; the validity of the linear, nonthreshold dose-response model; and risk as a basis for classifying hazardous wastes. As low as reasonably achievable (ALARA) is a radiation protection approach that keeps exposures to workers and the general public at levels as low as is practicable, taking into account social, technical, economic, practical, and public policy considerations. ICRP Reports 120 (1994), 127 (1998), and 129 (1999), as well as Statement 8 (1999) are focused on minimizing radiation exposures and applying ALARA principles. The basic principles of radiation protection, described in ICRP publications, form the basis of radiation protection legislation in most countries outside the United States. The NCRP in the United States provides recommendations, which closely follow those of the ICRP.

Justification of a radiation exposure requires an analysis of the radiation risk and of the benefit of the application [7]. Such an analysis must be carried out not only from a scientific or technical standpoint, it must also contain political, economic, and social aspects. In many countries, justification is mostly based on comparing the expected number of cases of damage due to the application of radiation with those caused by natural radiation exposure and other industrial effects.

As an example, 1 mSv y^{-1} over a working lifetime of 40 years could potentially expose a worker to an excess risk of fatal cancer of 0.16%, which assumes a nominal lifetime cancer mortality risk of 4% per Sv [13]. This was derived by extrapolation from high-dose rates and high doses (with the application of a dose/dose-rate reduction factor of two) to lower doses, where excess cancer risk cannot be identified statistically.

The external radiation hazard is controlled by applying the three principles: time, distance, and shielding. The dose accumulated by a person working in an area having a particular dose rate is directly proportional to the amount of time spent in this area. Limiting the time spent in the area can thus control the dose. Dose rates from about 0.1 μSv/h to a few tens of μSv/h are commonly encountered around facilities such as nuclear reactors. For a point source, the flux at a distance r is inversely proportional to the square of the distance r. For a point source, the radiation dose rate is directly related to the flux, therefore the dose rate also obeys the inverse square law. It can be noted that doubling the distance from the source reduces the dose rate to one-quarter of its original value, triplication of the distance reduces the dose rate to one-eighth, and so on.

The third method of controlling the external radiation hazard is by means of shielding. Generally, this is the preferred method because it results in intrinsically safe working conditions, while reliance on distance or time of exposure must involve continuous administrative control over workers. The amount of required shielding depends on the type of radiation, the activity of the source, and the dose rate that is acceptable outside the shielding material. One important problem, encountered when shielding against beta radiation, concerns the emission of secondary X-rays (bremsstrahlung), which result from the rapid slowing down of the beta particles. The fraction of beta energy, reappearing as bremsstrahlung, is approximately $ZE/3000$, where Z is the atomic number of the absorber and E is the β-energy in MeV. This means that beta shields should be constructed of materials of low mass number (e.g., aluminum or Perspex) to reduce the amount of bremsstrahlung emitted. A beta source emits beta rays with energies covering the complete spectrum from zero to a characteristic maximum energy, E_{max}. The mean beta energy is, in most cases, about $1/3 E_{max}$. The penetrating power of β-particles depends on their energy.

Gamma radiation and X-rays are attenuated exponentially when they pass through any material. The dose rate due to γ- or X-rays emerging from a shield can be written as:

$$D_t = D_0 e^{-\mu t} \qquad (2.11)$$

where D_0 is the dose rate without shielding, D_t is the dose rate after passing through a shield of thickness t, and μ is the linear absorption coefficient of the material of the shield. The linear absorption coefficient μ is a function of the type of material used for the shield and of the energy of the incident photons. It has the dimensions of (length)$^{-1}$ and is usually expressed in m^{-1} or mm^{-1}.

2.4 Dosimetry for Industrial Radiation Processing

The ability of ionizing radiation to induce effects that are potentially of use industrially has been known for many decades. However, it was not until the development of large-scale radiation sources, both radioisotopes and machines, in the 1950s that industrial scale radiation processing became a practical reality. Since that time, growth has been continuous and radiation processing now forms an important component of several industrial sectors (e.g., the polymer and medical device industries). One of the principal advantages of ionizing radiation as an industrial tool is the ability to achieve precise chemical and biological effects by the delivery of known doses of radiation. Accurate dosimetry has always been an important component of industrial radiation processing; the high doses and dose rates involved provide particular challenges. Currently there are three main applications for industrial radiation processing: the sterilization of medical devices, the treatment of foodstuffs, and the modification of polymers.

2.4.1 Medical Devices Sterilization

Approximately 50% of single-use medical devices are now sterilized by irradiation. The volume of irradiated products has been growing during the past years due, in part, to larger market demands for sterile medical devices and, in part, due to environmental concerns over the use of toxic gases, such as ethylene oxide, for sterilization. The volume of single-use medical devices now sterilized annually by ionizing radiation is estimated to be about 6 million cubic meters [14]. The majority of radiation-sterilized medical devices is currently irradiated by gamma rays from cobalt-60, but a steadily increasing fraction is irradiated by electron beams with energies of up to 10 MeV. There is also interest in the use of X-rays up to 5 MeV, and possibly higher, but so far only a few facilities are able to offer this kind of treatment. The upper energies for the use of electrons and X-rays are governed by the requirement to avoid induced radioactivity in the processed products. The minimum electron energy is determined by the penetration requirements, and can be as low as 50–100 keV for surface treatments. The validation and routine control of radiation sterilization relies on dosimetry. The minimum doses required for sterilization of medical devices are in the order of 10–30 kGy, but the actual dose depends on the regulatory requirement and on the level of initial microbiological contamination. Once the sterilization dose for a specific product has been determined, it must be delivered accurately. All products must receive at least the sterilization dose, but there is also a requirement to limit the maximum dose to the product in order to avoid deleterious effects on materials. The economics of the process will also be affected by an unnecessarily high dose. Taken together, these considerations lead to the need for tight dosimetric control of the process. As a part of the validation of an effective radiation sterilization process, it may be necessary to irradiate product samples with smaller doses in the range of 1–10 kGy. These doses do not achieve full sterilization but allow a determination of the effectiveness of the process. Dosimetric accuracy of the order of 5% to 10% is generally considered to be necessary for the effective control of the sterilization process. Dose is usually measured in terms of absorbed dose to water. Because of the implications for human health, the radiation sterilization process is under strict regulatory control. A number of international and national standards govern the practices and the required level of documentation [15, 16]. It is a common requirement that dosimetry is traceable to national standards and that the measurement uncertainty is known and documented.

In developed countries, radiation sterilization of disposable plastic medical items exceeds 50% of all such equipment, versus the older technique of chemical treatment with ethylene oxide. The percentage of such radiation-sterilized plastics continues to increase steadily [17, 18]. A major driver for the conversion to radiation sterilization is the elimination of toxic chemical residue from the product, as well as safety and environmental considerations gained in the sterilization process. A wide range of commercial medical products is now sterilized using either e-beam or gamma irradiation, including vials, tubing, gauze, sponges, syringes, and tissue-culture flasks [19]. A typical dose for radiation sterilization would be on the order of 2.5 kGy. The dose at which material properties are significantly affected varies tremendously among different polymer types and depends on irradiation conditions.

2.4.2 Food Irradiation

Food irradiation is the treatment of food by a certain type of energy. The process involves exposing the food, either packaged or in bulk, to carefully controlled amounts of ionizing radiation for a specific time to achieve certain desirable objectives. This can prevent the division of microorganisms, which cause food spoilage, such as bacteria and molds, by changing their molecular structure. It can also slow down ripening or maturation of certain fruits and vegetables by modifying/altering the physiological processes of the plant tissues. The Agricultural Research Service (ARS) has conducted a review of the scientific literature regarding the effectiveness of irradiation treatments for fruit flies in fresh fruits and vegetables with the goal of determining whether fruit fly generic dosages could be recommended [20]. Table 2.6 lists some examples of treatment levels and the desired effect on the food item.

The irradiation of foodstuffs is a valuable method for extending shelf life and reducing contamination by pathogenic microorganisms. The process has been used successfully in many countries for a number of years, although the total product volume is still relatively limited. Like the radiation sterilization process, the food irradiation process is highly regulated. The process and the documentation requirements for food irradiation are essentially the same as for radiation sterilization, and the same types of radiation sources are used. The doses ranges are, however, generally lower (0.1–15 kGy). Depending on the particular process or product, the requirements for dosimetric accuracy can be as strict as for radiation sterilization.

2.4.3 Modification of Polymers

It has been approximately 50 years since researchers first began exposing polymeric materials to ionizing radiation and reporting the occurrence of crosslinking

Table 2.6
Examples of Treatment Levels and the Desired Effects on the Food Item

Type of Food	Dose (kGy)	Effect
Meat, poultry, fish, shellfish, baked goods, prepared foods	20–70	Sterilization; treated product can be stored at room temperature
Spices and other seasonings	8–30	Reduces number of microorganisms and destroys insects
Meat, poultry, fish	1–10	Pasteurizes to protect against pathogens (salmonella, *e. coli*) and delays spoilage
Grain, fruit, vegetables	0.1–1	Kills insects or prevents them from reproducing; partially replaces fumigants
Bananas, avocados, mangos	0.25–0.35	Delays ripening
Pork	0.08–0.15	Inactivates trichinae
Potatoes, onions, garlic	0.05–0.15	Inhibits sprouting

and other useful effects [19]. The radiation technologies being applied to polymer processing comprise a diverse set, with many different radiation types and sources used, various types and combinations of materials, and many different application objectives being addressed [17, 21]. Crosslinking by radiation, a relatively mature technology, encompasses many product applications of significant economic impact. The industries utilizing radiation for producing crosslinked wire insulation and for heat-shrink products such as food wrap and tubing for electrical connections are particularly large. More recently, large-volume industrial production of self-resetting polymer-based electrical fuses and self-regulating heat tape has arisen using technology closely related to the heat-shrink products. Curing of coatings and inks is also well established and economically important for industry. Many other crosslinking applications that have been successfully commercialized are in the production of tubing, pipes, and automobile tires. More recently, radiation curing of fiber-matrix composites has begun to come into use.

The opposite of crosslinking—chain scission—is the basis of other radiation treatments aimed at enhancing processing characteristics of polymers. A large commercial industry exists in the irradiation of Teflon, which reduces particle size and molecular weight, and allows incorporation of the material into coatings, inks, and such [19]. Radiation sterilization of disposable plastic medical items (e.g., syringes, tubing, vials, gauze) has captured a large and still-growing segment of the market in many industrialized countries, due to advantages such as the elimination of toxic residues, which are problematic in chemical sterilization. As with a number of other radiation-processing technologies, control of unwanted material changes, including post-irradiation degradation in mechanical properties and discoloration, has been an integral part of commercialization and is still an active area.

The radiation treatment of low-value or waste agricultural byproducts based on naturally occurring polymers (including cellulose, starch, straw, and peanut shells), in an effort to produce useful products (e.g., polymer fillers, fibers, absorbents), remains an active area for the development in the processing technology. Surface modification and grafting of polymer chains, as well as modification of the bulk material, are employed in such treatments. The radiation treatment of consumer waste plastics for purposes of recycling is an emerging area of investigation motivated by both environmental and economic considerations [19].

Irradiation of polymers in order to modify material properties is the largest radiation process in terms of economic value. The achieved effects are crosslinking for heat-shrinkable products, increased melting points for cable insulation, and curing (polymerization) of coatings and inks. Some processes require tight dose specifications, but most are operated on the basis of achieved effects, rather than a specific dose measurement. The dose varies widely with the process and can be as high as 500 kGy. Electron irradiation is a predominant radiation source that has energies ranging from 0.1 to 10 MeV [22]. High-energy irradiation of polymers can convert them from dielectric materials to materials with higher electrical conductivity. This creates opportunities for use of radiation in producing specialty materials for electronic applications. Electrical conductivity of organic materials can be greatly increased during the time that the specimen is exposed to a radiation flux. This is due to the creation of transient conductive species (electrons, holes), which rapidly recombine once the irradiation is stopped. As a result, the conductivity is quickly decreased to near the initial value [23]. At a relatively high absorbed dose, permanent changes in the conductivity occur [19].

2.5 Medical Use of Ionizing Radiation

The use of ionizing radiation in medicine is justified and optimized and is based on the consideration of both the benefits and detriments of the exposure. Radiation used in medicine falls into two categories: diagnostic and therapeutic. X-rays are most often administered to the chest, limbs, and teeth. It is so frequent that it contributes significantly to the artificial background. Conventional radiography involves recording images on film or, increasingly, digital images stored on computer. Static images may be recorded or moving X-ray images can be viewed in real time on a display screen (fluoroscopy). Angiography is an application of X-rays for the examination of blood vessels and usually involves fluoroscopy. Angiography is used in a range of procedures, and the patient dose depends on the part of the body being investigated. Interventional procedures involve X-ray guidance for minimally invasive therapeutic procedures and are a combination of moving images that are viewed in real time on a display screen and spot-pictures that record static images. If the fluoroscopy is prolonged, these procedures give relatively high patient doses.

In computed tomography (CT), special X-ray equipment is used to take cross-sectional images of the body. The X-ray generator and detector rotate around the body, and computers construct two- or three-dimensional images of the body. Soft tissues are clearly visualized in CT images at the expense of relatively high patient doses.

In United Kingdom over the last few years, the frequency of CT examinations has increased by 39% and interventional radiology by 55%, while conventional radiology has only shown a 1% increase [24]. The sum of the average annual dose from nuclear medicine procedures, and an average of 0.38 mSv from X-ray examinations, results in average annual dose to the U.K. population of 0.41 mSv from all medical procedures.

Radiation therapy can be in external or internal form. In external therapy, the aim is to deliver as high a dose as possible to the tumor while limiting the dose to the surrounding normal tissue. High doses in the order of tens of Gy are needed to kill malignant cells. Whole body doses of these magnitudes would be fatal. The body can, however, tolerate high doses to local tissue volumes. With internal nuclear medicine, a radioactive source is administered to the patient. It could be in form of a drug or carrier material labeled with a radioactive material. The carrier is chosen so that it is preferably taken up by the malignant organ or tissue. The most frequently used radionuclide in nuclear medicine is technetium ^{99}Tc. Strict quality control procedures are essential for safe practice and disposal of these hazardous materials.

The decision to have an X-ray exam is a medical one, based on the likelihood of benefit from the test and the potential risk from radiation. For low dose examinations, usually those that involve only films taken by a technologist, this is generally an easy decision. For higher dose tests, such as CT scans and those involving the use of contrast materials (dyes), such as barium or iodine, the radiologist may want to consider the patient's past history of exposure to X-rays. If one has had frequent X-ray exams and changed healthcare providers, it is a good idea to keep a record of that patient's X-ray history. This can help the doctor make an informed decision.

With interventional radiology procedures using X-rays, the level of risk depends on the type of procedure. In general, the risk of developing a cancer from the exposure is not a major concern when compared to the benefits of the procedure. Many of the complex procedures, such as those used to open a partially blocked blood vessel, repair a weak area of a bulging vessel, or to redirect blood flow through malformed vessels, use extensive radiation. But such complex procedures are also frequently lifesaving in their benefit, and the risks associated with the radiation are a secondary consideration.

2.5.1 In Vivo Dosimetry

The accurate measurement of radiation dose during radiotherapy in the treatment of cancer requires knowledge of the actual dose delivered to internal organs with a highest level of accuracy. In addition to ensuring that the prescribed dose has been delivered to the cancerous tissue, it is also necessary to limit the dose burden to healthy tissue to as low as possible. The only way to ensure that critical organs are getting the dose required is to measure the dose inside the patient during treatment (i.e., in vivo dosimetry). Substantial progress has been made recently in this area. Development of an interactive, intraoperative dose planning system for seed implant brachytherapy in cancer treatment was reported [25]. This system involves in vivo dosimetry and the ability to determine implanted seed positions. It was achieved using a silicon minidetector, miniature front-end, and shaping amplifier with discriminator, connected to the minisilicon detector at the end of a cable placed in a urological catheter, to satisfy the spectroscopic requirements of the urethral probe. Metal oxide silicon field effect transistors (MOSFETs) were used as in vivo dosimetry detectors during electron beams at high dose-per-pulse intraoperative radiotherapy [26]. A novel linear array dosimeter consisting of a chain of semiconductors mounted on an ultrathin (50-μm-thick) flexible substrate and housed in an intracavitary catheter was designed for in vivo dosimetry during brachytherapy and diagnostic radiology [27]. The list of examples can be continued, as the advancements made in the area of in vivo dosimetry are impressive, and the research is still ongoing.

2.6 Uncontrolled Radioactive Releases

More than 400 nuclear power stations are in operation in more than 30 countries worldwide. France is the most dependent on nuclear power, generating about 70% of its electricity using nuclear fuels, with Belgium not very far behind [28]. Many of the other European countries generate between a quarter and half of their electricity this way, while the United States and the United Kingdom are about 20% dependent on nuclear power. Nuclear power plant accidents are the first in the following list of causes of nuclear radiation emergencies:

- Nuclear power plant accidents (Three Mile Island, Chernobyl).
- Nuclear materials processing plant accidents (Tokaimura, Japan).
- Nuclear waste (e.g., radioactive waste from hospitals, spent fuel and radioactive waste from nuclear power plants, radioactive contaminated materials) storage or processing facilities accidents.
- Nuclear waste transport truck or train accidents.
- Accidents involving nonwaste, but normal daily nuclear materials transport (i.e., trucks, planes, trains, couriers). One out every 50 shipments with hazardous materials contains radioactive substances. Approximately 3 million packages of radioactive material are shipped in the United States each year.
- Improper storage of radioactive materials (nonwaste) at any point during their normal material life cycle (e.g., power plants, medical, industrial, academic).
- Lost or stolen radioactive sources. (Over the last 50 years, incidents of lost and stolen licensed radioactive devices occur at the rate of once every other day [29].)

2.6.1 Nuclear Accidents

The International Nuclear Event Scale (INES) is a means for promptly communicating to the public in consistent terms the safety significance of events reported at nuclear installations. By putting events into proper perspective, the scale can ease common understanding among the nuclear community, the media, and the public. It was designed by an international group of experts convened jointly in 1989 by the International Atomic Energy Agency (IAEA) and the Nuclear Energy Agency (NEA) of the Organization for Economic Cooperation and Development. This scale is now operating successfully in more than 60 countries.

There are seven points in INES that grade nuclear accidents in terms of severity, shown in Table 2.7. Events from 1 to 3 are termed incidents, while those from 4 to 7 are accidents.

2.6.1.1 Examples of Rated Nuclear Events

The 1986 accident at the Chernobyl nuclear power plant in the Soviet Union (now in the Ukraine) had widespread environmental and human health effects. It is thus classified as Level 7. The 1957 accident at the Kyshtym reprocessing plant in the Soviet Union (now in Russia) led to a large off-site release. Emergency measures, including evacuation of the population, were taken to limit serious health effects (Level 6). The 1957 accident at the air-cooled graphite reactor pile at the Windscale (now Sellafield) facility in the United Kingdom involved an external release of radioactive fission products (Level 5). The 1979 accident at Three Mile Island in the United States resulted in a severely damaged reactor core. The off-site release of radioactivity was very limited (Level 5). The 1973 accident at Sellafield reprocessing plant in the United Kingdom involved a release of radioactive material into a plant operating area as a result of an exothermic reaction in a process vessel. It is classified as Level 4, based on the on-site impact.

Table 2.7
Grading Severity of Nuclear Accidents

7	Major accident	Major release with widespread health impact
6	Serious accident	Significant release
5	Accident with off-site effect	Limited release with severe core damage
4	Accident mainly in installation	Minor release with health effects to workers
3	Serious incident	On-site contamination
2	Incident	Potential safety consequences
1	Anomaly	Nonauthorized practice

In 1995, there were 17 events reported at Level 1, 8 at Level 2, and 1 at Level 3, while in 1996 there were 23 Level 1 anomalies reported in Sellafield [28].

2.6.2 Radioecology After Nuclear Accidents

2.6.2.1 Sellafield

British Nuclear Fuels (BNFL) complex at Sellafield in Cumbria, England, is involved in the storage and retreatment of reactor waste, not only from the United Kingdom, but from many foreign countries, among which Japan looms large [28]. Industry is permitted to legally pollute up to the regulatory limits. The Irish Sea is frequently referred to as the most radioactive in the world, due to the Sellafield discharges, most of which are licensed. In 1992, a leak of 30 liters of liquid plutonium caused the temporary closure of the plant. In 1986, half ton of uranium was released into the Irish Sea after a nuclear dump caused an increase in atmospheric radiation. In 1973, Sellafield had a problem with the corrosion of the metal cans containing the spent fuel rods, which led to a sharp increase in radioactivity levels [28].

2.6.2.2 Cap de la Hague

The Cap de la Hague center, located on the Normandy coast of France, commenced reprocessing spent fuel from commercial graphite/gas reactors in 1966. The complex was completed 10 years later by a high-activity oxide head-end facility to provide for reprocessing light water reactor (LWR) fuel. Considering routine discharges of low-level radioactive wastes from Cap de la Hague, one of the principal releases concerns liquid effluents discharged to the aquatic environment [30]. The release of liquid effluents is subject to very strict controls, with specification of maximum permissible activities. Various effluent treatment processes may be utilized (based on techniques such as co-precipitation and ion exchange) before discharge to the sea occurs. Statutory analyses are undertaken prior to the release of liquid wastes to assess the specific activity of radionuclides and the total activity of the release.

The absolute amount of activity discharged annually from Cap de La Hague is lower than that from Sellafield. It should also be noted that the composition of nuclides released from the two plants differs. At Cap de la Hague, besides tritium, the principal radionuclides released include ruthenium, cesium, strontium, and antimony. Routine discharges into coastal waters made by Cap de la Hague involved total releases of approximately 1.019×10^{16} Bq of ^{3}H; 4.905×10^{15} Bq of ^{106}Ru; 9.4×10^{14} Bq of ^{137}Cs; 7.55×10^{4} Bq of ^{90}Sr; 3×10^{12} Bq of ^{238}Pu + $^{239/240}$Pu; and 1×10^{15} Bq of ^{125}Sb [30]. A maximum in the release of ^{137}Cs from Cap de la Hague occurred in 1971, whereas a maximum release of ^{90}Sr occurred in 1983.

Discharges from the reprocessing operations of the Cap de la Hague plant give rise to contamination effects in the North Sea, contributing to concentrations of plutonium isotopes, americium, and curium. Regarding gaseous effluents, the dominant nuclide is ^{85}Kr, which gives rise to a dose of 3×10^{-4} Sv at the skin (for 800 tons of fuel reprocessed) within the maximal fallout area of Cap de la Hague's stack. In the terrestrial environment, it is considered that releases of gaseous effluents may give rise to contamination at the site, as a result of the passage of radionuclides into the subsoil. Therefore, the gaseous effluents released by the stack and the plant's dozen secondary outlets are monitored continuously.

2.6.2.3 Kyshtym

On September 29, 1957, a chemical explosion took place at Kyshtym in a storage tank containing 250 m^3 of high-level radioactive waste, generated as a result of plutonium production operations [30]. The explosion has been attributed to the ignition of an acetate-nitrate concentrate following the failure of a crude cooling system. This reportedly has led to the release of 7.4×10^{17} Bq of activity to the atmosphere. While most of this release was deposited on the ground close to the site of the explosion, a plume of finer particulates (containing about 10% of the activity released) was carried to a height of 1 km and in the form of a radioactive cloud was transported to the north and northeast. The largest contribution to the total activity of the release (accounting for over 60%) was due to ^{144}Ce + ^{144}Pr. Additional contributions were from ^{95}Zr + ^{95}Nb (over 20%), ^{90}Sr + ^{90}Y (over 5%), ^{106}Ru + ^{106}Rh (over 3%), and ^{137}Cs (over 0.03%) [30]. Detection of ^{89}Sr, ^{147}Pm, ^{155}Eu, and plutonium in the release has also been reported. The principal source of radiation dose to biological surfaces over the first year after the accident was ^{144}Ce + ^{144}Pr, while ^{90}Sr was the principal contributor to long-term exposure. The resulting terrestrial distribution of radioactivity may be considered to comprise two phases associated with initial radioactive fallout and wind migration. Deposition of virtually all radioactivity from the plume occurred within 11 hours along a 300-km path from Kyshtym, leading to the contamination of an area that extended over 2×10^4 km^2. Effects of wind resuspension with regard to redistribution of contaminants were most apparent during the first few days after the accident but did not lead to significant changes in the trace boundaries. By 1958, wind transfer caused the redistribution of less than 1% of the original radioactive fallout, the formation of the radioactive trace being essentially completed.

Biogeochemical processes, such as soil vertical migration and leaching, subsequently affected the distribution of radioactive fallout in the environment. In the spring following the accident, most activity (90% to 95%) was concentrated in the turf, with 0.5% to 1.5% in living plants and 5% to 10% in the mineralized portion of the soil [30].

As a result of the release, all pine trees within 20 km^2 in which the dose to needles exceeded 30–40 Gy perished by late 1959. The loss of farm animals exhibiting acute radiation sickness symptoms was reported to have begun within 9–12 days in sites nearest the accident. With regard to the aquatic environment, half periods for the elimination of radionuclides from lake water varied from 1 to 24 days for ^{144}Ce, and from 780 to 1,100 days for ^{90}Sr. Contamination densities in highly contaminated rivers ranged from 4×10^3 to 28×10^3 times higher than preaccident levels immediately after the accident [30]. Over a 25-year period, the ^{90}Sr concentration in lake water within the contaminated region decreased by a factor of 30. In total, almost 11,000 people were evacuated from a 700-km^2 area in which contamination exceeded 74–148 GBq km^{-2} ^{90}Sr. Additional measures taken to mitigate the circumstances of the accident included decontamination of portions of agricultural land, monitoring of agricultural produce, introduction of restrictions on the use of contaminated areas, reorganization of agriculture, and forestry. The 170-km^2 area that was most highly contaminated following the accident has remained unsuitable for human habitation, agriculture, or forestry and is designated for research purposes as a radioecological reserve [30].

Although average external exposure doses preceding evacuation are reported to have reached 0.17 Sv, with effective dose equivalents of 0.52 Sv (1.5 Sv to the gastrointestinal tract), it should be noted that significant uncertainties are associated with these figures, which may be doubled to account for the nonuniformity of contamination density and exposure conditions. It should be mentioned as well that at the time of the accident, there were no dosimetry devices available for civilian use. Inhabitants had to completely rely on the government authorities and their recommendations (if any). Delayed alert of nuclear emergencies resulted in considerable health and life losses among the population, much of which could have been prevented. Later sections of this chapter discuss commercially available and emerging affordable personal radiation dosimeters (see also Chapters 4 and 5), which could make a difference in situations like this.

2.6.2.4 Three Mile Island

The first nuclear power plant accident occurred in 1979 at the U.S. Three Mile Island Unit 2 (TMI-2) in Pennsylvania and was attributed to the combined effects of equipment malfunctions and operator error. The amount of radioactivity released into the environment was of the order of 10^{17} Bq and consisted mainly of the noble gases ^{133}Xe, ^{133m}Xe, and ^{135}Xe [30]. Additionally, around 1.11 TBq of ^{131}I was released in the course of the accident. Environmental monitoring undertaken in the locality of TMI following the accident included analysis of samples of milk, air, water, soil, vegetation, fish, and river silt and sediment. Increased environmental radionuclide concentrations were observed only due to ^{131}I in cow's and goat's milk, on-site nondrinking water and air, and ^{137}Cs in fish.

The highest doses received as a consequence of this accident were considered to be to people living within a 3.2-km radius of the plant. An estimated dose of 0.2–0.7 Sv was received by 260 people, mostly located on the east bank of the local river.

2.6.2.5 Chernobyl

The most serious accident to have occurred in the history of nuclear reactor operation was on April 26, 1986, at the Unit 4 reactor of the Chernobyl nuclear power plant located in the Ukraine. An uncontrollable growth in reactor power led to fuel rupture and succeeding explosions. Ejection of material containing spent fuel, enriched with noble gases and volatile nuclides of iodine, tellurium, and cesium, into the atmosphere occurred during the initial explosions, with a continuing release due to the subsequent fire in the graphite moderator. Efforts to suppress radionuclide releases and to prevent further fission involved covering the core with more than 5×10^3 tons of boron, dolomite, sand, clay, and lead. This caused the emissions to be decreased substantially for a few days; however, the blanketing of the core led to increased temperatures and a further release peak from May 1 to 5. Emissions were terminated on May 6 upon pumping a nitrogen coolant through tunnels constructed under the core [30].

At the end of the 10-day release period following the accident, some 2×10^{18} Bq of activity was present in the environment. Besides a total release of noble gases, some 20% of the iodine available in the core was released (6.7×10^{17} Bq), with around 10% of the inventory of cesium (1.9×10^{16} Bq ^{134}Cs, 3.7×10^{16} Bq ^{137}Cs) and about 3% of rare earths and actinides. Of the fraction of the core ejected to the atmosphere, 0.3% to 0.5% was deposited on the site, 1.5% to 2% landed within 20 km, and the remainder was dispersed beyond 20 km. The release spread across Europe from Poland. A section turned northward to the United Kingdom; other material traveled eastward across the Soviet Union and southward to Turkey and Greece. By mid-May, levels of the order of 1×10^{-3} Bq m^{-3} were observed in Hong Kong and North America [30]. The observed contamination pattern of the deposition of radioactive material was often concentrated in a few small areas within countries and is highly correlated with local variations in amounts of rainfall during the passage of the plume.

As a consequence of this accident, a 30-km radius zone around the power plant was evacuated, and an area of more than 2×10^4 km^2 outside this zone has been designated as contaminated with Cs, Sr, Pu, Zr, Nb, Ru, and La isotopes. Agricultural land comprised about a quarter of the contaminated territory; therefore, food production was restricted. Various measures to reduce contamination at the reactor site were also implemented, including such activities as collection and disposal of loose fuel fragments, removal of topsoil, application of special solutions to effect decontamination, and the containment of the damaged Unit 4 within a concrete *sarcophagus*.

The total amount of radioactivity in the river water in the vicinity of the accident was calculated to be 7.4×10^{15} Bq. It was concluded that these levels did not exceed prescribed limits acceptable for drinking.

In the Chernobyl accident, three people were killed almost immediately; about 400 people, including firemen, were involved in the initial response to the explosion and received the highest radiation doses. About 240 people were hospitalized with severe radiation sickness and 28 of them died within weeks of accident. Their doses ranged from 1 to 16 Sv [28]. The majority of those with doses higher than 5 Sv died. Official figures claim that 31 died as a result of the accident. Over 800,000 people participated in the liquidation of the consequences of the accident. Their average dose is estimated to be about 100 mSv, while 10% received doses of the order of 250 mSv. Many of the liquidators received external doses comparable with Japanese atomic bomb survivors, and similar health consequences—leukemia, cancer, and immune system damage—have occurred with passage of time [28]. Thyroid cancer is the most obvious and scientifically accepted health effect of the accident. With a 10-day delay from the first explosion, over 100,000 local people were evacuated from a 30-km exclusion zone around the power plant. Hundreds of thousands have died of cancer in the region since the accident, but the number attributable to radiation from the accident is difficult to determine, given the absence of personal dosimeters for civilian use at that time.

2.6.3 Further Examples of Nuclear Contamination

A list of recent incidents with stolen or lost radioactive sources can be found at http://www.ki4u.com/lost.htm. Nuclear contamination of packages, workers, and delivery vehicles has occurred, and one could receive a package that has been contaminated by a leaking source. Radioactive sources are being lost, stolen, or illegally discarded every other day. A postal transporter jet was grounded until a radiation survey could determine whether or not it was contaminated by a faulty shipment of radioactive material, which had already irradiated other equipment, a vehicle, and a person. A popular radioactive industrial gauge is stolen at the rate of once a month. For example, a doctor in Indiana, Pennsylvania, disregards radiation alarms after a procedure, which leads to the irradiation of more than 90 persons and kills the patient. In 1996, the *Harrisburg Patriot News* reported that a woman near Pittsburgh took into her home a contaminated floor scrubber that had radioactive levels more than 1,000 times more than the levels considered safe [31].

Cockpit and cabin crew members who spend hundreds of hours a month flying are exposed to a much higher level of cosmic radiation than the general public. There are four principal factors that affect the increased radiation dose received by those who fly a lot: altitude, latitude, hours aloft, and solar activity. On average, the amount of cosmic radiation roughly doubles with every 2,000m increase in altitude. At a typical long-haul airline cruising altitude of FL390, cosmic radiation is around 60 times greater than at sea level. For an intercontinental flying at FL510, the dose is nearly 200 times greater than at sea level. The Earth's magnetic field causes most of cosmic radiation to be concentrated near the north and south magnetic poles. The exposure rate at 70° north or south latitude is about four times as much as at 25°. Thus, flights over polar routes (e.g., New York to Tokyo or Chicago to London) are exposed to a lot more radiation than those confined to the mid-latitudes.

Considering numerous nuclear accidents described in Sections 2.6.1 and 2.6.2, and such reasons as lost radiation sources, transportation of nuclear waste, radiological terrorism, and the possibility of nuclear weapons being used in a war, the members of the public should be encouraged to use personal electronic dosimeters. They will not, of course, evaluate internal doses, but would provide real-time alerts if the safe limit of the external doses are exceeded. For a long time, radiation dosimetry was the privilege of nuclear specialists, military personnel, and medical personnel. Unfortunately radiation is now used in many areas and quite often affects civilians. The future of dosimetry is dividing in two parallel paths that depend on application: high-precision expensive dosimetry for medical and space purposes, and affordable safeguard wide-range real-time dosimetry, where targeted device price is not more than $20.

2.7 Review of the Principles and Materials in Radiation Dosimetry

2.7.1 Dosimetry

Radiation dosimetry is the process of determining the energy absorbed in a specified target from a radiation field. The processes by which energy is transferred from the radiation field and absorbed in the target tissue depend on the nature and energy of the radiation. The general principle underlying most methods of detection of nuclear radiation is that whatever the form of the radiation, it gives up part or all of its energy to the detecting medium either by ionizing it directly or by causing the emission of a particle, which in turn produces ionization in the medium. The ionization produced is then detected by one of an increasing variety of techniques, discussed next. When an energetic nuclear particle penetrates a semiconductor, it knocks many atoms out of their normal lattice position by its impact [32]. Most of these knocked-out atoms find their way back to normal lattice positions, but there are always a few lattice sites left vacant and perhaps a few atoms left in interstitial positions. If either of these lattice imperfections is capable of trapping either holes or electrons, it will have a marked effect on the electrical conductivity of the material. In germanium of moderate resistivity, it is observed that irradiation with high-energy particles always seems to produce electrically active p-type centers. This means that p-type material will lower its resistivity, while n-type material will tend to convert to p-type [32]. The predominant effect of radiation on organic solids is to displace atoms or ions, producing defects in the lattice structure. These defects are responsible for coloration of the sample and for changes in their physical properties, particularly electrical conduction. A lattice vacancy will be produced when an atom or ion is displaced and interstitial defect will be formed (if the atom or ion is trapped between the lattice planes). The total energy requirement for vacancies plus an interstitial species is about 25 eV. Although only 1 eV is required to form the vacancy and 4 eV for the interstitial species, another 20 eV is required for the disruption of the neighbors. These defects may be produced by direct displacement or as a result of ionization.

An electron can be readily trapped in an anion vacancy; this is known as F-center. It causes coloration and can be detected by electron spin resonance (ESR). The ESR spectra can provide information on the interaction of the trapped electron with the neighboring ions, and frequently several types of F-centers can be distinguished. These color centers and other radiation-induced defects in materials may be removed by annealing or ultraviolet light. Considerable annealing takes place with only a small temperature rise after irradiation at low temperatures, such as near a few degrees Kelvin (K). At certain temperatures corresponding to defect or trap energies, strong luminescence is frequently observed.

The choice of dosimetry system for a particular application is usually based on a number of factors, such as the nature and intensity of the radiation to be measured, the time available for counting, the efficiency, and the cost. The sensitivity of a particular system is usually expressed as a *lower limit of detection* (LLD), which is also referred as *minimum detectable dose* (MDD) or *minimum detectable amount* (MDA).

The LLD for which the risks of false negative results and of false positive results are each 5% is defined as:

$$LLD = 4.66(SD_b)/E_{ff} \quad (2.12)$$

where 4.66 represents the product of the distribution parameters needed to establish the 5% error limits, E_{ff} is the detection efficiency, expressed in counts/dis, and SD_b is standard deviation of the background count according to:

$$SD_b = N_B^{0.5} = (R_b \times T)^{0.5} \quad (2.13)$$

where N_b denotes total background counts in time T, R_b is the background count rate, and T is the minimum counting time required (in minutes).

The main emphasis must be on selecting a system with the highest possible efficiency E_{ff}. The requirements of low-level counting systems are high counting efficiencies, low background count rates, and high stability, since long counting periods are often necessary.

2.7.2 Classification and Calibration of Dosimeters

In 1998 the International Electrotechnical Commission published its first standard IEC 61526 to be applied to personal dosimeters [33]. These are worn on the trunk of the body and are used for the measurement of personal dose equivalents, or personal dose equivalent rates received by the wearer. This standard covers physical characteristics of the dosimeter (ideal version); its radiological, mechanical, and environmental performance; safety settings; and reliability of readout system. For example, dosimeter mass should not exceed 200g, with maximum dimensions of $15 \times 3 \times 8$ cm^3 excluding clip or other fixing arrangement. The dosimeter shall withstand drops from heights of 1.5m onto a hard-tiled surface, without affecting its response to radiation, within $\pm 10\%$. The dosimeter response shall not vary more than $\pm 20\%$ relative to 20°C in the range $-10°C$ to 40°C and than $\pm 50\%$ in the range $-20°C$ to 50°C.

No single instrument at present has all the characteristics described. Accordingly, different types of instruments must be used, depending upon the nature of the radiation hazard and the different dosimetry application domains where radiation is generally involved. The classification of dosimeters can be done as well following such criteria as the nature of involved physical phenomena, characteristics of the measured fields, or the types of applications (e.g., [34]):

- Radiation protection (external dosimetry);
- External radiation therapy;
- Radio diagnostic (two-dimensional and three-dimensional) imaging: low doses (a few mGy);
- Industrial applications.

The passive dosimetry mode corresponds to the use of a dosimeter, where certain physical characteristics are modified by the incident radiation. Such a dosimeter integrates the dose, and the measurement is performed after a time delay using a specific reading device. TLDs, radiological films, nuclear emulsions, and many others proceed from the same method. Imaging X-ray and γ-ray detectors with large areas, high detection efficiencies, and excellent spatial resolutions over broad X-ray energy ranges have applications in nondestructive testing, astronomy, medical imaging, macromolecular crystallography, and basic research [35]. Film radiography has long been used as a principal imaging method for these applications. Although it provides superior spatial resolution, this method is inefficient, extremely time consuming, labor intensive, and unsuitable for real-time applications. Modern and more sophisticated digital X-ray imaging systems are based on combinations of a series of scintillating phosphor screens coupled to the charge-coupled device (CCD) or amorphous silicon detector arrays (a-Si:H). This combination offers the potential for very high spatial resolution, dynamic range, and a wide range of system formats that can be easily modified to meet specific application requirements. The principal advantages of CCDs for personal dosimetry are twofold: high sensitivity at low dose rates due to low-noise operation and a wide dynamic range. It was proposed that the differing properties of the two halves of the sensor increase its dose range [36]. Deposition of scintillator coatings of various thicknesses across the surface of the sensor provides sensitivity to a wider energy range.

Other radiation measurement methods include:

- Pulse counting for low dose rate: photons and electrons in radiation protection;
- Pulse counting, energy, and time discrimination: mixed fields in radiation protection;
- Direct current measurement: high dose rate of photons and electrons;
- Measurement of electrical characteristics of passive dosimeters (photons, electrons, and neutrons): change in electrical resistance, drift in threshold voltage, change in capacitance, and so forth.

The International Organization for Standardization (ISO) has proposed a system of classification of sealed radioactive sources based on the safety requirements for typical uses (ISO 2919:1999). This system provides a manufacturer of sealed radioactive sources with a set of tests to evaluate the safety of his or her product. It also helps a user of such sealed sources to select types that suit the application in mind and accordingly to select the proper detector for radiation monitoring.

2.7.2.1 Calibration

Many types of ionizing radiations are employed in a variety of useful and steadily increasing applications. This progress in radiation technology has necessitated the development of scientific methodology aimed at the protection of workers and the public from potential harm. Some uses of radiation technology have critical requirements for accurate dosimetry, because they are directly related to health and safety. For example, radiation therapy for the treatment of cancer and other diseases requires extremely accurate measurement of the dose delivered to a tumor volume and an equally accurate assessment of the dose delivered to surrounding normal tissues in order to ensure the success of the therapy. The dosimetric accuracy required for radiation protection purposes is somewhat relaxed compared to radiation therapy; nevertheless, accurate dosimetry is also needed in this field [37]. Industrial radiation processing, such as the sterilization of disposable medical products, also relies on accurate dosimetry to ensure adequate sterilization while not delivering an excessive dose that might destroy the product being treated. Most dosimeters exhibit dose response that is dependent on the energy of the radiation measured, so that corrections are nearly always applied to the readings of dosimeters and survey meters to determine the required dosimetric quantity.

Calibration is the set of operations that under specified conditions establish the relationships between values indicated by a measuring instrument or values represented by a material measure and the corresponding known values of a measurement. Complexity of the calibration procedure depends on what information is required and extends from a full type to routine tests. In full type test, a device is exposed to a range of radiation energies and angles of incidence as well as other influence parameters. In routine tests, calibration factors are established or confirmed for individual personal dosimeters or batch samples. Reference radiations are normally maintained at national calibration laboratories. It is essential that a radiation-measuring instrument, or dosimeter, calibrated at any laboratory yield the same result if it was calibrated at any other qualified metrology laboratory. An impressive number of national and international standards have been developed and published. These documents not only carefully describe methods for performing accurate calibrations, but also contain large amounts of relevant physical data necessary for the interpretation of dosimetric measurements and the assessment of associated uncertainties.

The international standard that describes the requirements for operation of testing and calibration laboratories is ISO 17025 [38]. This standard outlines the critical elements needed for proper operation of a calibration laboratory, including personnel, laboratory environment, equipment, procedures, records, quality system, measurement traceability, and reporting of results. Conformity to the specifications of this standard is usually required for laboratories seeking certification or accreditation [37]. A series of standards specific to particular reference radiations is available and includes photons [39–42], beta particles [43, 44], and neutrons [45–47].

In the ISO 11137 standard, the photons generated by ^{137}Cs or ^{60}Co sources are stated as reference photon energies for the response of a personal dosimeter [15]. Special care should be taken when calibrating the devices at low energies, where the photoelectric effect is predominant for solid-state detectors, leading to a strong over-response [34].

The detector sensitivity is a function of the dose: a higher energy of electrons and photons during irradiation gives a larger sensitivity drop for the same dose. One of the major problems associated with high-dose dosimeter calibration is the effect that environmental factors, such as temperature, gases, dose rate, and humidity, can have on the response of the dosimeter. It is necessary to state under what conditions a calibration was carried out in order for corrections to be made subsequently if the dosimeter is used to measure dose in different conditions. In the case of routine dosimeters, it is necessary to calibrate the dosimeter under the conditions of final use. Consequently, some form of subsequent verification of the calibration has to be carried out. This often involves the use of reference dosimeters to check the validity of the routine dosimeter calibration at a number of dose points in the plant of final use.

Only a very few dosimeters exhibit a strict linear relationship between the readout signal and absorbed dose. This means that it is generally not possible to define a single calibration factor for a dosimeter, and a nonlinear calibration function has to be used instead [14].

2.7.3 Gas-Filled Detectors

Gas-filled detectors consist of a cylindrical cathode with a window, an axial wire anode, and a sensitive volume of gas. The gas may be continuously replenished, giving a flow through the detector, or the detector may be sealed. In the high electric field surrounding the anode wire, electron multiplication occurs through a process of gas amplification. The *gain* of this process is defined as the number of electrons collected on the anode wire for each primary electron produced in the original ionizing event.

According to the potential applied to the anode, the detector can work as an ionization chamber, proportional counter, or Geiger counter. An ionization chamber is a gas-filled detector without any gas amplification. In proportional counters, the electric potential is high enough for the gain to reach a value in the range from 10^2 to 10^5. Each electron produced by the initial photo-ionization causes one avalanche. Since the number of avalanche events is proportional to the energy of the incident photons, the charge collected by the anode is proportional to the incident photon energy. The proportional gas scintillation counter consists of a proportional counter coupled to an ultraviolet-sensitive photomultiplier tube. Initial electrons produced by the interaction of the high-energy photon with the counter fill-gas are accelerated by a high electric field, where they acquire sufficient energy to excite the noble gas atoms. The resulting ultraviolet (UV) radiation is observed by a photomultiplier tube. The statistical uncertainties in the number of formed ion-electron pairs and gas gain result in a pulse amplitude distribution, where photons of constant energy are absorbed. Detector resolution Γ, is a measure of its ability to resolve two peaks that are close together in energy. Γ (in eV) is the full width at half maximum (FWHM) of its pulse amplitude distribution, given empirically by:

$$\Gamma \cong 2.35\sqrt{F\varepsilon E} \qquad (2.14)$$

where F is the Fano factor, whose value is a property of the detector and is empirically determined; E is the photon energy in eV; and ε is the ionization energy. For example, for argon gas detector $\varepsilon = 15.8$ eV, $F = 1.7$, and for 6,000 eV photons, Γ reaches the value of around 950 eV.

The Fano factor F parameterizes the fluctuations in the number of ion pairs and is defined by:

$$F = \sigma_e^2 / N_e \qquad (2.15)$$

where σ_e^2 is the variance of the charge expressed in units of the electron charge e and N_e is the number of ion pairs. $F = 1$ corresponds to Poisson statistics. Fano originally predicted that the charge fluctuations would be sub-Poissonian, $F < 1$, because the individual ion-pair creation processes are not independent once the additional constraint $E = N_e W_e$ is included, where E is the energy deposited by the incident particle and W_e is the mean energy required to create a free electron-ion pair.

In Geiger counters, gas amplification reaches saturation, and proportionality no longer exists. The output signal does not depend on the incident energy. The time taken for the counter to recover from the saturation is called *deadtime*.

2.7.3.1 Ionization Chambers

Ionization chambers measure dose and dose rate from gamma and x-radiations. Ion chambers come in a variety of chamber volumes, according to the sensitivity required. A typical ionization chamber that measures total absorbed dose is the pocket dosimeter, which is the size of a large fountain pen. It has a chamber containing two electrodes, one of which is a quartz fiber loop free to move with respect to its mounting. Radiation entering the chamber causes ionization and excitation of gas molecules along its passing track [48]. The distance to which the fiber moves is proportional to the dose received in the chamber. The great advantage of this type of instrument is that it can be read at any time without the aid of a supplementary charger-reader by simply holding it up to a source of light and looking into it.

Chipmunk ion chamber Model 1055 available from Far West Technology (http://www.fwt.com) is a three-terminal 3.4-liter ionization chamber designed for measuring low levels of radiation.

An example of a commercially available pen-sized ionization chamber is FH 39 Pocket Dosimeter from Thermo Electron Corporation (http://www.thermo.com). FH 39 dosimeters are available in three versions that differ in their energy range as shown in Table 2.8.

2.7.3.2 Geiger-Mueller Counters

Geiger-Mueller (GM) counters are normally used for detecting single ionizing events, which take place within the sensitive volume of the counter. They are very rugged and sensitive to low levels of radiation. They are usually equipped with audible as well as analog and digital output. GM counters can detect gamma photons or alpha and beta particles. Detection of gamma rays is less efficient than of beta particles. A discriminating shield is usually provided with GM instruments, which when opened admits both beta and gamma radiation. With the shield closed, only gamma is admitted. Use of the shield may permit qualitative differentiation between ionization caused by beta particles and that produced by gamma photons.

Examples of GM Counters Model GCA-04 Digital Geiger Counter with RS-232 Serial Output available from Images SI (http://www.imagesco.com) takes real-time radiation readings and displays the count/second and equivalent radiation level in either mR/hr or mSv/hr. Detector sensitivity is above 3.0 MeV for alpha, above 50 keV for beta, and above 7 keV for gamma radiation (http://www.imagesco.com/catalog/geiger/digital_counter.html).

PM1203M Programmable Dosimeter (Figure 2.11) provided by Polimaster (http://www.polimaster.com) is a GM tube with equivalent dose rate measurement in the range of 0.1 μSv/h to 2,000 μSv/h. The response energy range is 0.0–61.5 MeV, with no more than 10 seconds of response time. The minimum detectable dose of PM1203M is 0.01 μSv.

PM1604A is a compact personal dosimeter (Figure 2.12) also provided by Polimaster. It uses a GM tube as radiation detector and is capable of measuring the personal dose equivalent $H_p(10)$ and personal dose equivalent rate for both gamma and X-ray radiation from 1 μSv/h up to 10 Sv/h in the energy range

Table 2.8
FH 39 Pocket Dosimeters

Model	Measuring Range	Energy Range
FH 39 UP	0 ... 2 mSv	45 keV ... 3 MeV
FH 39 EP	0 ... 50 mSv	45 keV ... 3 MeV
FH 39 RP	0 ... 2 mSv	18 keV ... 3 MeV

Figure 2.11 PM1203M Programmable Dosimeter. (Courtesy of Polimaster, Inc.)

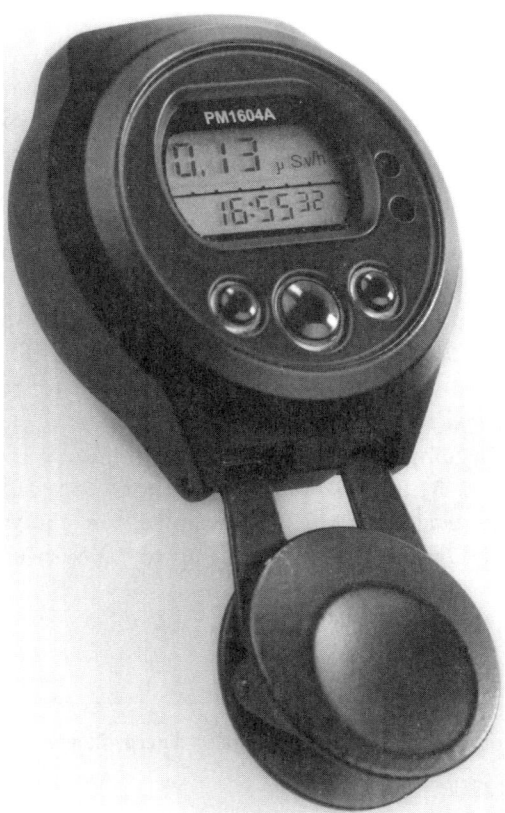

Figure 2.12 PM1604A Compact Personal Dosimeter. (Courtesy of Polimaster, Inc.)

0.048 MeV–6.0 MeV. The dosimeters enable the users to preset two independent dose and dose rate alarm thresholds, and when these thresholds are exceeded the dosimeter gives audible alarm signals. The minimum detectable dose of PM1604A is 0.01 μSv. The dosimeter stores up to 1,000 histories of dose rate measurements, accumulated dose values, events, and levels of the preset alarm thresholds exceeding in their nonvolatile memory. This information can be transmitted to a personal computer through infrared channel using special software for further processing and analysis.

PM1621A is another example of a personal dosimeter that uses a GM tube, also provided by Polimaster. Its lowest limit of detection is 0.01 μSv in the energy range of 10.0 keV–20.0 MeV. It was designed as a professional dosimeter to measure personal dose equivalent and personal dose equivalent rate of both gamma and X-rays. Its high sensitivity enables the PM1621A to register even the slightest variations of the natural background.

The PM1208 Wrist Gamma Indicator is a unique watch (the first dosimeter watch in the market) that continuously monitors the ambient dose and dose rate as well as the irradiation level of the person wearing the indicator (Figure 2.13). Equally designed for use by professionals dealing with radiation on a daily basis and for use by concerned citizens, the PM1208 alerts users to hazardous situations through an audible alarm. Polimaster additionally recommends this instrument for all frequent airplane travelers to monitor their radiation exposure.

Figure 2.13 The PM1208 Wrist Gamma Indicator. (Courtesy of Polimaster, Inc.)

2.7.3.3 Proportional Counters

The proportional counter is a type of a gas-filled detector that operates in a pulse mode and relies on the phenomenon of a gas multiplication [48]. Unlike the Geiger counter, the pulse size in the proportional counter is related to energy deposited in the detector. Therefore, it is possible to distinguish the larger pulses produced by alpha particles from the smaller pulses produced by betas or gamma rays. Generally, proportional counters are used to detect one type of radiation in the presence of other types of radiation or to obtain output signals greater than those obtainable with ionization chambers of equivalent size. Proportional counters may be used to either detect events or to measure absorbed energy (dose), because the output pulse is directly proportional to the energy released in the sensitive volume of the counter. Proportional counters are most widely used for the detection of alpha particles, neutrons, and protons. Since in the proportional counter the electron must reach the gas ionization level, there is a threshold voltage after which the avalanche process occurs. In typical gases at atmospheric pressure, the threshold field level is on the order of 10^6 V/m.

In general, the proportional gas should not contain electronegative components, such as oxygen. Otherwise, electrons heading toward the anode would combine with the electronegative gas. If this happens, a negative ion goes to the anode and will fail to produce an avalanche. The result is that the pulse is probably too small to exceed the threshold setting and be counted. Air is sometimes used as a proportional gas for alpha counting, but it could not serve for beta detection, as beta particles produce far fewer ion pairs in the gas than alphas. Using air as the proportional gas allows the use of a thin window without the need for a gas flow system. However, it is essential that the air be dry. In high-humidity conditions, air proportional counters are prone to generating spurious pulses. The fill gas in a proportional counter (and a GM detector) is usually a noble gas, since noble gases are not electronegative and do not react chemically with the detector components. Of the noble gases, argon is the most widely used because of its low cost. Other noble gases with higher atomic numbers (e.g., krypton and xenon) might be used if increased sensitivity to X-rays or gamma rays is required. Hydrocarbon gases (e.g., methane, propane, and ethylene) can also serve as a fill gas, but they have the disadvantage of being flammable. For certain applications in dosimetry, it is desirable that the detector has the same type of response as human tissue to radiation. To accomplish this, a tissue-equivalent gas mixture, such as 64.4% methane, 32.4% carbon dioxide, and 3.2% nitrogen, might be used.

He-3 and BF_3 are the most commonly employed gases in neutron detectors. These gases serve a dual purpose. First, thermal neutrons undergo nuclear reactions with the He-3 or BF_3 to produce charged particles. Second, the charged particles ionize the He-3 or BF_3 to produce pulses. An example is Model 42-30, a proportional neutron detector from Ludlum Measurements (http://www.ludlums.com). It is a form of a wall-mountable ball and is used for area monitoring in energy range from thermal to approximately 12 MeV. Complete technical specifications are available at http://www.ludlums.com/product/m42-30.htm.

2.7.4 Scintillation Counters

A scintillation counter combines a photomultiplier tube (PMT) with a scintillating material. The operating principle of these detectors is based on the ability of certain materials to convert nuclear radiation into light. The most widely used scintillators include the inorganic alkali halide crystals—such as sodium iodide (NaI), cesium iodide (CsI), zinc sulfide (ZnS), and lithium iodide (LiI), of which sodium iodine is the favorite—and organic-based liquids and plastics [48]. Bismuth germanate ($Bi_4Ge_3O_{12}$), commonly referred to as BGO, is used in applications where its high gamma counting efficiency or its lower neutron sensitivity outweigh considerations of energy resolution.

When gamma rays interact with the scintillator material, ionized (excited) atoms shift to a lower energy state and emit photons of light. In a pure inorganic scintillator crystal, the return of the atom to lower energy states with the emission of a photon is an inefficient process. Furthermore, the emitted photons are usually too high in energy to lie in the range of wavelengths to which a photomultiplier tube is sensitive. Small amounts of impurities (called activators) are added to all scintillators to enhance the emission of visible photons. Crystal de-excitations channeled through these impurities give rise to photons that can activate the PMT. One important consequence of luminescence through activator impurities is that the bulk scintillator crystal is transparent to the scintillation light. A common example of scintillator activation encountered in gamma-ray measurements is thallium-doped sodium iodide NaI(Tl).

Scintillation photons incident on the photocathode, liberate electrons through the photoelectric effect, and these photoelectrons are then accelerated by a strong electric field in the PMT. As these photoelectrons are accelerated, they collide with electrodes in the tube, releasing additional electrons. This increased electron flux is then further accelerated to collide with the succeeding electrodes, causing a large multiplication (by a factor of 10^4 or more) of the electron flux from its initial value at the photocathode surface. Finally, the amplified charge arrives at the output electrode (the anode) of the tube. The magnitude of this charge is proportional to the initial amount of charge liberated at the photocathode of the PMT. The constant of proportionality is the gain of the PMT. Furthermore, the initial number of photoelectrons liberated at the photocathode is proportional to the amount of light incident on the phototube, which, in turn, is proportional to the amount of energy deposited by radiation in the scintillator (assuming no light loss from the scintillator volume). Thus, an output signal is produced that is proportional to the energy deposited by the radiation in the scintillation medium. However, the spectrum of deposited energies (even for a monoenergetic photon flux) is quite varied, due to the occurrence of the photoelectric effect, Compton effect, and various scattering phenomena in the scintillation medium and statistical fluctuations associated with all of these processes.

Scintillation counters that are available can detect alpha and beta particles, gamma rays, neutrons, protons, and electrons. Although very energy dependent, scintillation counters are more efficient at detecting low-level gamma backgrounds than GM counters. Numerous examples of commercially available scintillator counters with detailed technical characteristics can be acquired from relevant manufacturers (see the Appendix).

2.7.4.1 Novel Developments

Novel materials and material compositions with enhanced properties are being developed, which lead to constant improvement in the efficiency of scintillation counters. A wide range of both organic and inorganic materials is being implemented for dosimetry purposes, and extensive research work is ongoing. A few examples are discussed next.

In recent years, halide perovskites of the ABX_3-type (e.g., $KMgF_3$ and $LiBaF_3$) have become the object for luminescence study, laser media, and radiation detectors development. The variety of crystal nomenclature allows one to select different energy gaps, whereas the crystal lattice gives prospects doping as compared to simple (two-component) alkali and alkali-earth halides. Therefore, variation of the perovskite composition, selection of proper dopants, and thermal treatment conditions form a basis for radiation sensor materials development. To modify emission parameters, $KMgF_3$ and $LiBaF_3$ crystals were doped with Ce^{3+} ions [49]. The investigations showed that this activator allows one to shift the spectrum of UV radiation toward the region convenient for registration, decrease the contribution of the delayed exciton luminescence, and partially bind the residual oxygen anions.

Due to fast and worldwide introduction of β-sources for intravascular application, there is a growing interest in the dosimetry aspects. However, accurate dosimetry of β-radiation is more difficult than that of γ-radiation. Water-equivalent plastic scintillators are considered suitable detectors for this application [50]. Plastic scintillators can be used as dosimeters because they emit visible light proportional to the absorbed dose rate. A scintillator does not disturb the radiation field due to its mass absorption coefficient and its mass stopping power, which is water-equivalent in a wide range of energies. Furthermore, plastic scintillation dosimeters provide a fast and direct reading of the measuring values combined with a high spatial resolution. Depending on the spatial resolution needed, the scintillators can be as small as 1 mm^3 or less. Radiation-induced changes in properties of three plastic scintillators, namely, BC-408, EJ-200, and

BC-404, are described in [51]. An example of parallel read-out system with plastic scintillators for dosimetry in medical applications is given in [52].

2.7.4.2 RadDetect PRD 1250 Personal Radiation Detector

One of the most compact real-time personal dosimeters on the market at the moment is RadDetect PRD 1250 Personal Radiation Detector (Justacip, http://www.RadDetect.com), shown in Figure 2.14. The RadDetect has a dual-mode sensor, which converts directly and indirectly incident radiation into an electric signal.

When a radioactive photon strikes a depletion region created by reverse bias on the photodiode, it produces a small amount of charge in proportion to the photon's energy (direct conversion). The resulting signal is then amplified and processed by the central processing unit (CPU) in the RadDetect. A high-grade scintillator material in an indirect conversion–type detector converts incoming radiation energy first into the visible light. The visible light is then captured by the photodiode and converted to an electric signal (indirect conversion). The resulting signal is then amplified and processed by the CPU in the RadDetect.

According to manufacturer's specifications, RadDetect PRD 1250 Personal Radiation Detector detects β- and γ-radiation, X-rays, and fast neutron radiation in dose range from 75 mR/h to 2,000 R/h (Justacip, http://www.RadDetect.com). Detection ability includes radioactive iodine (^{131}I) from a nuclear reactor accident or fallout from nuclear bomb as well as dirty bomb–type radiation such as ^{137}Cs, ^{90}Sr, and ^{60}Co. Radioactive iodine is a by-product of nuclear fission processes in nuclear reactors and nuclear weapons. One of the benefits of the real-time auto-alert (both audible and visual) capability of the RadDetect is that the user is able to monitor increasing/decreasing

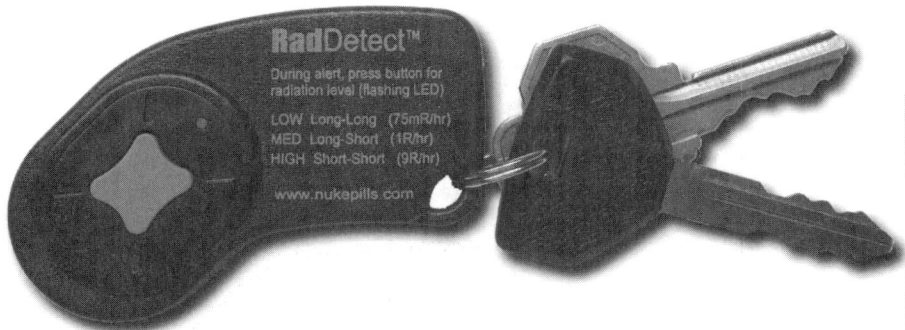

Figure 2.14 RadDetect PRD 1250 Personal Radiation Detector. (Courtesy of Justacip, Inc.)

radiation rate, allowing him to determine the direction from which radiation is coming. Recommended applications include:

- Alert first responders to radioactive threats;
- Use by customs and border patrols;
- Use by law enforcement;
- Use by security officers in nuclear power facilities, banks, government laboratories, and medical facilities;
- Use by the military;
- Use by government agencies;
- Use by hazmat teams;
- Use by fire departments;
- Civilian use for personal protection.

Gamma-Neutron Pager PM1703GN The PM1703GN provided by Polimaster (http://www.polimaster.com) is the world's first unique gamma/neutron pager of the new generation, capable of detecting the least amounts of radioactive and nuclear materials, including the weapon ones (see Figure 2.15). PM1703GN uses two independent detectors to detect gamma and neutron radiation separately: CsI(Tl) scintillator with photodiode for gamma detection and $Li^6I(Eu)$ with photodiode for neutron radiation. PM1703GN energy range is 0.033–3.0 MeV for gamma rays and 0.025 eV–14 MeV for neutrons. The algorithm

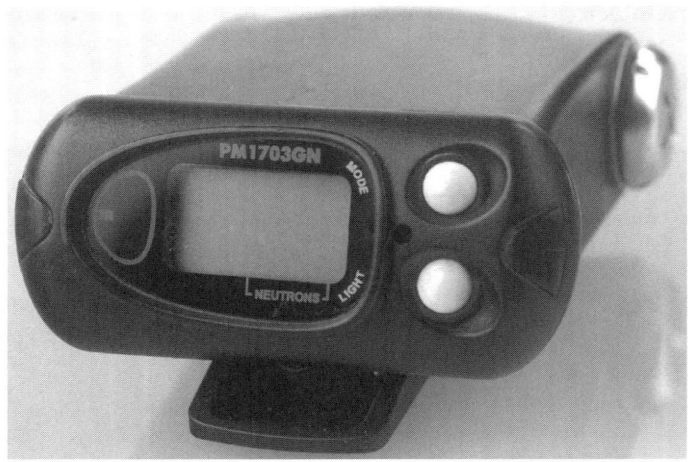

Figure 2.15 Gamma-Neutron Pager PM1703GN. (Courtesy of Polimaster, Inc.)

implemented in the instrument enables the user to change the value of the background, so that the instrument can be used both as the alarming device for controlling the variations of radiation level in the controlled area and as the effective instrument for searching and locating the source in the event of alarm of the fixed installed radiation monitors. The nonvolatile memory of the instrument makes it possible to store the history of the instrument's operation for its further transmission to a PC through the built-in or external IRDA adapter for database forming.

2.7.5 Chemical Dosimeters

Chemical dosimeters are systems in which measurable chemical changes are produced by ionizing radiation. Many molecules dissociate by radiation, but only a few can be used as dosimeters, mainly due to calibration problems. The reactions taking place in chemical dosimetry have two steps, a very fast ($10^{-11}-10^{-10}$s) ionization by interaction of energetic charged particles and a chemical reaction according to the chemical kinetics. Radiation produces acids in the system, the amount of which can be determined from visible color changes or, more accurately, by titration or pH readings. Most chemical systems of practical size are useful only for gamma doses of hundreds to millions of Gy.

The Fricke dosimeter, also known as ferrous sulphate dosimeter, is an optical density sensor. It is secondary standard chemical dosimeter based on the change in oxidation state from ferrous ions to ferric ions [53]. The Fricke dosimeter consists of an aerated dilute solution of ferrous sulfate in 0.8 N sulfuric acid: 0.001 M $FeSO_4$ or $Fe(NH_4)_2(SO_4)_2$ + 0.8 NH_2SO_4 air saturated [54]. This solution is glass-sealed, and the absorbance of light is read by a spectrophotometer. The maximum absorbed radiation dose is reached when all ferrous ions are converted into ferric ions.

2.7.6 Radiochromic Dye Films

Radiochromic effects involve the direct coloration of a material by the absorption of energetic radiation without requiring latent chemical, optical, or thermal development or amplification. Radiochromic films based on polydiacetylene, known as GafChromic MD-55(I and II), are widely used for medical applications. The film consists of a thin microcrystalline monomeric dispersion coated on a flexible polyester film base. The film is clear (translucent) before it is irradiated. It turns progressively blue upon exposure to radiation. The radiochromic radiation chemical mechanism is a relatively slow polymerization reaction initiated by irradiation, resulting in homogenous, planar polyconjugation along the carbon-chain backbone. The increase in absorbance is roughly proportional to the absorbed dose of ionizing radiation. The bluing becomes relatively stable after 24 hours. The response of the film is spatially nonuniform and depends on coating.

Radiochromic film has many important limitations that must be considered when using it to determine the dose distribution in any measurement. There is a marked difference in the response with temperature fluctuations; therefore, temperature during calibration and experimentation should be the same. Most radiochromic films are sensitive to ultraviolet radiation, which spontaneously colors the film. Sensible care must be taken to protect the film from sunlight or white fluorescent lights. There is negligible dependence on changes in humidity. The films should always be kept in a dry dark environment at the temperature and humidity at which they will be utilized. The film orientation and alignment should be consistent during actual use and scanning to determine dose distribution, as the sensitive layers have a preferred direction when irradiated.

Commercial radiochromic dye films have been used in recent years to quantify absorbed dose in several medical applications. For example, the GafChromic MD-55-2 dye film was found suitable for photon irradiation in brachytherapy [55].

GEX Corporation (http://www.gexcorporation.com) manufactures prepackaged radiochromic film dosimeters, which are available in various configurations.

2.7.7 TLDs

2.7.7.1 Theory of TLD Gamma Dose Response

Luminescence For all nonmetals, energy injected into a material appropriately makes it luminesce, and the energizer can be a photon, a particle, or even an impact. Most nonmetals possess energy levels, which can store a quantum of electronic energy and then re-emit it as a photon, usually in the visible region, such as energy 2–3 eV (blue-green) [2]. In some cases, the electronic energy is stored until heat or light is used to stimulate the dosimeter material. Phosphor is known as an efficient material to convert incoming beams into blue-green quanta.

Detection of Radiation with Thermoluminescent Materials Many crystalline materials exhibit thermoluminescence. Light is emitted when the material is heated after the exposure to radiation. Electron and hole traps are filled during radiation exposure (solid-state ionization effect); heating frees electrons and holes from these traps. Light is emitted when the electrons and holes recombine. The emitted light is a characteristic of the material and a function of temperature. The plot of light output versus temperature is known as the glow curve for that particular material. A schematic *glow curve* of phosphor material is shown in Figure 2.16.

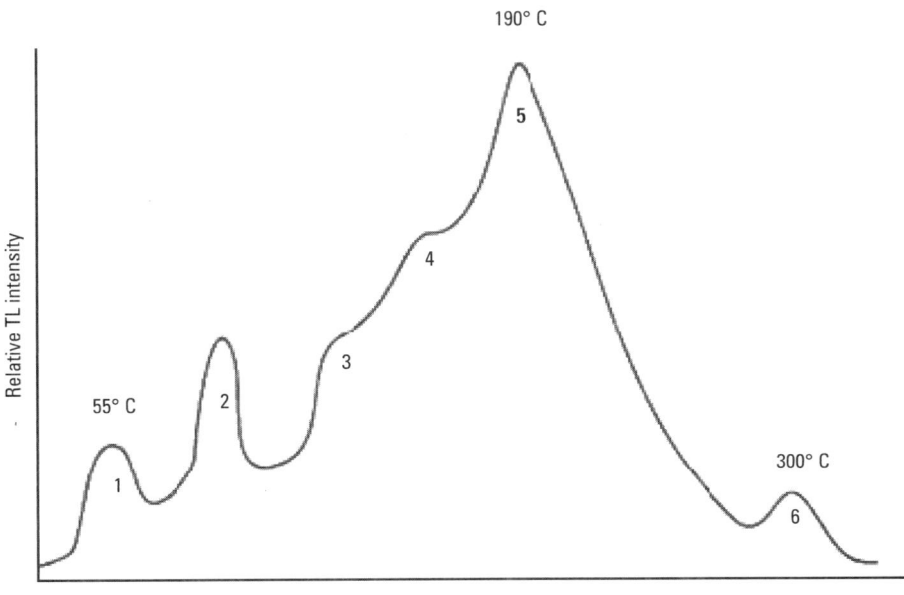

Figure 2.16 A schematic glow curve of phosphor material.

Usually the highest peak is used to calculate the dose equivalent. The area under the curve represents the radiation energy deposited on the TLD (assuming linearity between TL light output and the radiation dose), and the glow curve may be taken as a fingerprint in TL dosimetry. TLD is widely used for personal dosimeters as well as for area monitoring. Examples of TLD materials are compounds of LiF, $Li_2B_4O_7$, BeO, CaF_2, and $CaSO_4$, doped with various activators. LiF, $Li_2B_4O_7$, and BeO are nearly tissue-equivalent for photons, while CaF_2 and $CaSO_4$ have the advantage of being 20 to 30 times more sensitive. The smallest dose that can be measured with TLDs is in the range of few mGy. The maximum dose is generally quoted as being in the range of $10^2 - 10^3$ Gy, but as these levels are approached there is some loss of linearity [2]. Repeated exposures to high dose levels may result in a permanent loss of sensitivity. TLDs are subject to fading, which is quite small in the case of LiF (few percent over a period of months), but can be quite high in the case of $CaSO_4$ and CaF_2 (10% to 50% in a matter of hours). TLDs can be made sensitive to neutrons as well as gamma rays; they also exhibit energy dependence according to the materials. Therefore, with careful selection and combination of TLDs, and the use of different shielding materials, it is possible to discriminate between types of radiation.

As mentioned earlier, thermoluminescent material detects radiations by the formation of a meta-stable crystalline structure with the valence electrons.

When lithium fluoride absorbs the radiation, it raises electrons in the crystal to higher energy levels. Some of these electrons are trapped by impurities, where they remain in their excited states until the crystal is heated. When heated, the electrons are released from these traps and return to their lowest energy state with the release of the amount of light, proportional to the radiation exposure. Once the crystal is heated to a sufficiently high temperature, all the trapped electrons are released and the dosimeter may be reused.

When alkali halide crystals, such as lithium fluoride LiF, which appear transparent in nature, are subject to an appropriate chemical or physical treatment, they acquire a distinctive color. The coloring of the crystals results from distortions of the crystalline structure. A color center (F center) is formed when the dislocation of a negatively charged ion in the crystal lattice leaves an electron trapped in the positively charged vacancy [56]. An F center can be visualized as an electron confined in a hole in the crystal lattice, much as an electron is confined in a hydrogen atom, giving origin to absorption bands in the visible range of the spectrum responsible for the coloring of the host crystal. The clustering of these defects leads to the formation of aggregate F centers, which can be visualized as hydrogen molecules with their own distinctive absorption spectra. LiF is unique among the alkali halides because only such physical process as irradiation can color it. When the irradiation is performed using X- and γ-rays, the defects are formed uniformly throughout the sample, and the measurement of the absorption spectra when the defect concentration is high could be difficult. On the other hand, the coloring performed with low penetrating radiations (low energy electrons, ion beams) causes the formation of defects in a thin layer at the surface of the sample. This simplifies the recording of the absorption bands, which remain measurable even for high irradiation dose values.

The color center concentration is derived from the absorption band peak heights by applying Dexter's modification to the Smakula equation:

$$Nf = \left(0.87 \times 10^{17}\right) \cdot \frac{n}{\left(n^2 + 2\right)^2} \alpha_{max} W \qquad (2.16)$$

where f refers to the oscillator strength, n is the refraction index at the absorption band peak wavelength, W is the absorption band half-width expressed in eV, and α_{max} is the absorption coefficient at the band peak expressed in cm^{-1}. An important feature of the coloring curves is the evident saturation of the center formation, beyond which the growth trend is reversed and the defect concentration starts declining. This behavior is common to all alkali halides [57] and thermoluminescent materials [58].

The dose response of most materials used in thermoluminescence dosimetry usually shows a linear, then supralinear, then saturating response [59]. At very high dose levels, damage effects can appear, which reduce the TL signal as the dose is increased. The normalized TL dose response $f(D)$ is a measure of the relative TL efficiency. In the linear dose region $f(D) = 1$; in the region of supralinear dose response $f(D) > 1$; and at very high dose, where damage effects are dominant, $f(D) < 1$. One of the features of TL dose response lies in the dependence of the supralinearilty on energy and type of radiation.

The complexity of the TL mechanism arises from the multistage nature of the conversion from the initial energy imparted by the radiation field to the final energy liberated as TL photons. This conversion occurs in three stages [59].

1. Absorption of energy from the radiation field via the capture of charge carriers at defect trapping centers, accompanied by possible creation or alteration of radiation-induced defects, which participate or compete with the TL process. The various defects, which have captured a charge-carrier of either sign, can serve as trapping centers (TCs), luminescent centers (LCs), or competitive centers (CCs).
2. Absorption of thermal energy in nonisothermal annealing, which liberates the charge carriers and can bring about changes in the trapping structures and their spatial correlations.
3. Dissipation of the thermal and radiant energy via diffusion of the charge carriers through the crystal lattice, followed by recombination, in which a certain fraction of this energy results in the formation of TL photons. Recombination can occur via charge carrier diffusion in the conduction band, which can be viewed as a delocalized or long-range mechanism, or via recombination between a locally trapped electron-hole pair. This mechanism gives rise to the linear/low-dose part of the response.

To understand in more detail the theory of thermally and optically stimulated luminescence phenomena in materials, one may refer to [59–62].

2.7.7.2 Photographic Emulsions, TLD, and OSLs

Photographic emulsions are frequently used as detectors. The film badge has been the most common dosimeter in use but is tending to be replaced by thermoluminescent dosimeters. Film badges are used for nonreal-time monitoring of personal exposure to beta and gamma radiation. To assess various radiations simultaneously, a strip of film is covered with absorbers. Some film badges have a small window shielded by a sheet of mylar, which can detect beta radiation, and one or more sections shielded by metal foils for detecting gamma radiation. The radiation exposure of the film is determined by the degree of darkening of the film after it is developed. TLDs measure the amount of energy (dose) per unit mass of material absorbed by the material when exposed to ionizing radiation. The TLD material absorbs and stores energy when exposed to radiation, and heating the TLD releases the energy as light. The advantages of this method are that they are of a small size, relatively inexpensive, have integrated measurement, and are reusable. The main disadvantage is that the TLD requires evaluation through a reader (e.g., noninstantaneous readout). Some TLDs are sensitive enough to measure a dose of β- or γ-radiation of a few μGy, and some TLDs can also detect neutrons. The example of a film badge that uses the effect of radiation on photographic film to record dose is shown in Figure 2.17. After film developing, the optical density is compared to a film calibration curve, and a measure of exposure dose is obtained. As the exposure dose increases, the optical density of the emulsion increases. At least two different types of films are employed to cover a wide exposure range: a low-exposure film, 0.2 mGy to 20 mGy, and a high-exposure film, 10 mGy to 10 Gy. Metal filters such as aluminum, copper, and cadmium-tungsten are used to provide information on radiation quality—to differentiate between different radiation types. The heavy metal filter also intensifies the gamma radiation interaction. Beta radiation is evaluated by observing the density change to a portion of the film, which is not covered by a filter. Film badges or TLDs are widely used, as they provide an accurate means of recording radiation exposure at a low cost. Their disadvantage is that heat, moisture, and aging will cause a natural change in the film's optical density.

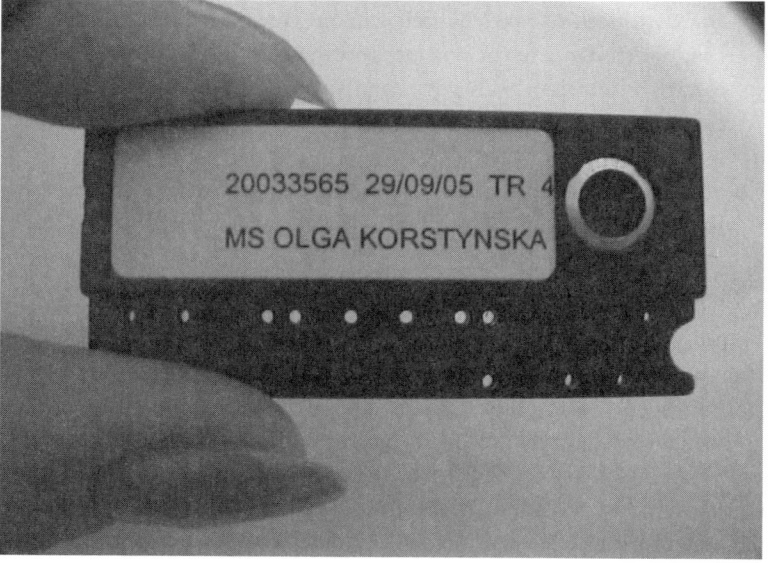

Figure 2.17 The film badge dosimeter.

OSL dosimeter technology uses a laser to stimulate luminescence in a thin layer of aluminum oxide material, which was exposed to radiation. The amount of exposure can be determined by shining a green light on the crystal and measuring the intensity of the blue light emitted. OSL systems allow instantaneous readings that can be repeated, as opposed to TLDs, which take 20 or 30 seconds for a one-time-only reading.

Radiophotoluminescent (RPL) glass is a dosimeter material that will luminesce following an excitation pulse of ultraviolet light, if it has been exposed to ionizing radiation. This effect is caused by radiation-induced changes in the glass crystalline electronic structure. Although other materials also exhibit this property, silver activated RPL glass has found the greatest application in X-ray and γ-radiation dosimetry. The sensitivity depends on the type and manufacturer selected, and ranges from 0.1 mGy to several million mGy. This type of dosimeter cannot be zeroed; it gives a total cumulative dose reading that fades very slowly with time.

TLD manufacturing differs from company to company, so specific chip arrangement and composition may vary. Most badges are lithium fluoride (LiF) and calcium fluoride (CaF). Lithium has two stable isotopes, ^{6}Li (sensitive to neutrons) and ^{7}Li. Badges that measure betas and gammas have at least one chip behind a mylar window to allow some energy discrimination of betas and soft X-rays.

Neutron dosimetry is often made with the TLD-700 chip, which is made with ^{7}LiF, sensitive to betas and gammas. The TLD-600 chip is made with ^{6}LiF, which is sensitive to betas, gamma, and neutrons. The neutron dose is calculated from the difference between a TLD-600 and TLD-700 pair. Some cards use four TLD chips, arranged as two pairs, to measure neutron dose. One TLD-600 and one TLD-700 pair are shielded from the front with cadmium (Cd), which absorbs neutrons. A second pair is shielded with Cd from the rear. The readings of these four TLD chips are combined into an overall calculation of neutron dose.

The most common TLD badge in the commercial nuclear power industry is the Panasonic UD-802 badge [63], which is capable of estimating dose received at the three tissue depths (7 mg/cm^2, 300 mg/cm^2, and 1,000 mg/cm^2) that are specified in regulations for reporting shallow, lens of eye, and deep dose equivalent. With four independent detection elements, the badge can measure dose from beta, gamma, X-ray, or neutron radiation over a wide range of energies. The badge may be used for monitoring personnel in medical, industrial, and other nuclear applications. It is calibrated using ^{137}Cs gamma rays and may be used for routine radiation monitoring of gamma or X-rays over the energy range from 30 keV to 1.25 MeV. The badge is calibrated over the beta energy range between Tl-204 (0.267 MeV) and Sr-90/Y-90 (0.565 MeV). Composition of Panasonic UD-802 Dosimeter is given in Table 2.9.

Table 2.9
Composition of Panasonic UD-802 Dosimeter

UD-802	Element 1	Element 2	Element 3	Element 4
Phosphor	$Li_2B_4O_7$	$Li_2B_4O_7$	$CaSO_4$	$CaSO_4$
Shielding	Plastic	Plastic	Plastic	Plastic and lead
Thickness	18 mg/cm^2	360 mg/cm^2	360 mg/cm^2	1,040 mg/cm^2

The thin plastic shielding of element 1 allows beta radiation to penetrate to the $Li_2B_4O_7$ phosphor to induce a response. The plastic shielding over elements 2 and 3 is at a depth that high-energy beta radiations penetrate to the phosphor. The $CaSO_4$ phosphor over-responds to low-energy photon radiation. The plastic over element 3 allows low energy photons to penetrate through and cause response in that phosphor. The lead filter over element 4 attenuates the low-energy photons, thus reducing the intensity of the photon radiation that reaches the phosphor and causes a response.

Minimum measurable dose (see Section 2.7.1) for Panasonic UD-802 Dosimeter was experimentally determined as 3 mGy for $CaSO_4$:Tm and 20 μGy for $Li_2B_4O_7$:Cu [63]. The low fading characteristics of calcium sulphate phosphor (2% to 9% over 90 days) make this dosimeter a good choice for environmental monitoring.

The dosimetric properties of the KLT-300 (KAERI LiF:Mg,Cu,Na,Si TL detector) TL detector were investigated [64]. The sensitivity of the TL detector was about 30 times higher than that of the TLD-100. For comparison, Figure 2.18 shows the typical glow curves of LiF:Mg,Cu,Na,Si TL detector and of the TLD-100. The dose response was linear from 10^{-4} to 10 Gy (see Figure 2.19), and a sublinear response was observed at higher doses. The photon energy response of the detector from 20 to 662 keV showed maximum response of 1.004 at 53 keV and minimum response of 0.825 at 20 keV. A detection threshold of the detector was found to be 70 nGy [64].

The Luxel's OSL dosimeter from Landauer (http://landaueriii.com) measures radiation exposure due to X-ray, beta, and gamma radiation through a thin layer of aluminum oxide doped with carbon. Al_2O_3:C has a TL sensitivity some 50 times greater than that of the industrial standard TLD material (TLD-100 [LiF:Mg,Ti]) and is almost tissue-equivalent. However, it possesses the undesirable properties of a strong sensitivity to light and thermal quenching. This aluminum oxide is stimulated with the use of a laser light, which causes the aluminum oxide to become luminescent in proportion to the amount of radiation. The badge is designed to measure radiation exposure in the range of 1 mRem to

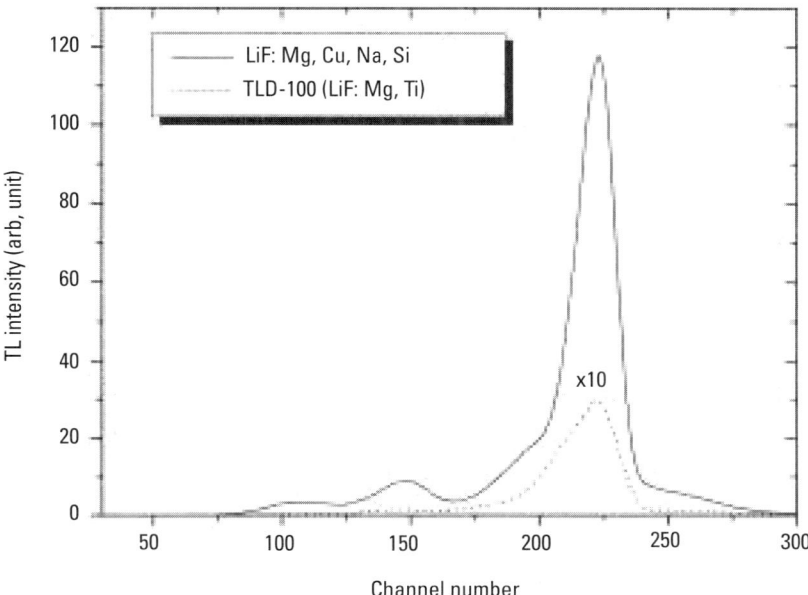

Figure 2.18 Typical glow curves of LiF:Mg,Cu,Na,Si TL detector and of the TLD-100 for the comparison of sensitivity. (*From:* [64]. © 2004 Elsevier. Reprinted with permission.)

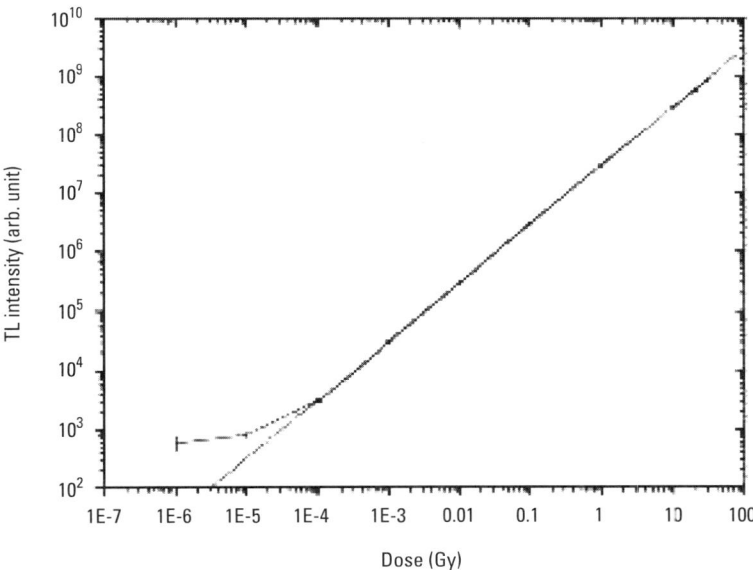

Figure 2.19 Dose response of LiF:Mg,Cu,Na,Si TL detector as a function of absorbed dose. (*From:* [64]. © 2004 Elsevier. Reprinted with permission.)

1,000 Rem for x- and γ-radiation; and 10 mRem to 1,000 Rem for beta radiation. The dose is recorded as whole body dose, and the badge can be reread to confirm the accuracy of a radiation dose.

2.7.8 Radiation Dosimetry Using MOS Devices

MOS devices are used for integrated ionizing dose measurements. They exploit the fact that charge generation and trapping in the gate insulator of a MOS device will cause a shift in electrical characteristics, notably the threshold voltage, and this shift is proportional to the radiation dose [65]. In MOS devices, the amount of charge trapped depends strongly upon the voltage across the oxide during irradiation. A secondary effect (e.g., production of new interface states and slow-state instabilities) involves the rearrangement of atomic bonds at the oxide-silicon interface and is contributing to the device degradation. A detailed study on radiation effect in MOS structures can be found in [66].

The electrical consequences of radiation-induced physical changes in MOS devices are progressive loss of function and eventual failure of an MOS circuit [65]. The threshold voltage of an MOS transistor is ascertained by measuring the channel current (drain current), I_D, as a function of the gate voltage, V_G, at a constant supply voltage, V_{DD}. The characteristic curve for an n-type channel in a p-type substrate is shown in Figure 2.20(b). The minimum in the C-V curve, Figure 2.20(a) reflects the transition from depletion to inversion conditions. Comparing Figure 2.20(a) and Figure 2.20(b), it is seen that the value marked as V_T on the I-V characteristic is always aligned with V_i, a point near this minimum at which inversion is established [65]. The shifts observed in irradiated MOS devices can be divided into two largely independent elements: the voltage shifts caused by oxide-trapped charge and that caused by interface-trapped charge. Trapped charge causes simple translation of the I-V characteristic. Interface charge causes shift and distortion of the I-V characteristic.

The radiation-sensing field-effect transistor (RADFET) is an integrating radiation dosimeter based on silicon technology, which stores the dosimetric information in the form of a trapped hole charge for a very long time unless specially erased. Dose readings remain in the RADFET after it has been electronically interrogated to determine the total ionizing radiation exposure. Compared with conventional dosimeters, RADFET has many advantages, which include minute sensor size, low cost, entirely electrical sensor-reader interface, operation at low voltages and currents, and dynamic range from millirads to megarads. The RADFET can be designed and configured to monitor ionizing radiation for a broad range of applications, such as performing radiation dose measurements without applied electrical power or remote sensing in locations with limited access. The RADFET provides a direct electrical voltage output and has a simple control circuit that can be integrated on the same small chip with the sensor.

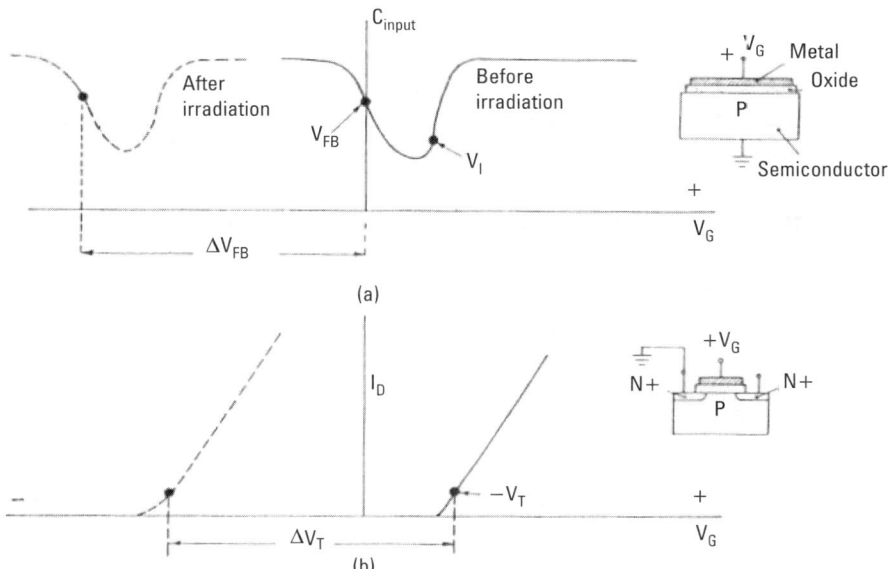

Figure 2.20 MOS characteristics: typical variation of (a) capacitance and (b) drain current with gate voltage, showing the shifts in flatband and threshold voltages due to trapped charge. (*From:* [65]. © 1993 Oxford University Press. Reprinted with permission.)

Exposure to ionizing radiation—such as gamma rays, X-rays, electrons, and high-energy protons—causes the RADFET's voltage output to change in a predictable manner. The construction of a RADFET sensor is that of a conventional p-channel MOS transistor. The response of this FET sensor is the change in threshold voltage ΔV_T, or the shift in flatband voltage ΔV_{FB}, as a function of integrated radiation dose. Its sensitivity to ionizing radiation is dependent on the dosimeter's gate-dielectric thickness and on the magnitude of its gate-biasing voltage during irradiation [65]. When constructing the sensor, the designer manipulates the values of oxide thickness, which is usually larger than that needed for normal FETs to achieve high response. The effects of other trapping regions near the interfaces of the oxide are reduced as much as possible in the fabrication of the dosimeter.

2.7.9 Interaction of Ionizing Radiation with Semiconductor Detectors

Historically, semiconductor detectors were conceived as solid-state ionization chambers. Conduction counters (highly insulating diamond crystals) were first used for the detection and possibly spectroscopy of the ionizing radiation. However, such crystals were quickly rejected because of poor charge collection characteristics resulting from the deep trapping centers in their band gap. After the highly successful development of silicon (Si) and germanium (Ge) single crystals for transistor technologies, the conduction-counter concept was abandoned. Silicon and germanium ionizing radiation detectors were developed by forming rectifying junctions on these materials. The use of silicon diode detectors for dosimetry of high-energy photon and electron beams, mainly in the field of radiation therapy, began by the mid-1960s [34]. A semiconductor detector was a large silicon or germanium diode of the p-n or p-i-n type operated in the reverse bias mode. At a suitable operating temperature (normally $\sim$300K for silicon detectors and $\sim$85K for germanium detectors), the barrier created at the junction reduces the leakage current to acceptably low values. Thus, an electric field can be applied that is sufficient to collect the charge carriers liberated by the ionizing radiation.

Silicon diodes, solid-state ionization chambers made with silicon dioxide or diamond, and transistors (mainly MOSFETs) are mostly used for radiation dosimetry. Following the type of dosimetry, semiconductor devices can be used either as electronic or real-time dose rate meters or as solid-state dosimeters. In the first case, the involved dosimetric quantity corresponds to dose rate and is correlated with the produced electric current. Dose is obtained by integration. In the second case, the dosimetric quantity corresponds to the absorbed dose and is correlated with the modification of a given physical parameter, such as electric conductivity for the diode or threshold voltage for the MOSFET. Mean dose rate can be obtained by means of sequential dose measurements. For these two types of dosimeters, the solid-state effects that occur often coexist. This means that a progressive decrease of sensitivity occurs during the increase in the absorbed dose. In general, the result is a saturation effect [34].

2.7.9.1 Heavy Charged Particles

Heavy charged particles lose energy by Coulomb interaction with the electrons and the nuclei of the absorbing materials. The collision of heavy charged particles with free and bound electrons results in the ionization or excitation of the absorbing atom, whereas the interaction with nuclei leads only to a Rutherford scattering between two types of nuclei. Thus, the energy spent by the particle in electronic collisions results in the creation of electron-hole pairs, whereas the energy spent in nuclear collisions is lost to the detection process.

Polycrystalline ZnS(Ag) was the first solid material employed as a particle detector, and it is still the most widely used scintillator for gross alpha counting. For high sensitivity, the source thickness should never be permitted to exceed the range of the alpha particles (see Section 2.1.2). Discs of transparent plastic coated with silver-activated ZnS and photomultiplier tubes coupled to appropriate electronic circuitry remain in regular use for environmental monitoring. The background count rates over the entire alpha-energy range are relatively low and of the order of one event per hour for ZnS(Ag) on a thin plastic disc of 24-mm diameter.

In the case of alpha-particle spectrometry, maximum energy resolution is achieved with a source, which approximates as closely as possible to that of an infinitely thin, weightless source on a perfectly flat substrate. A wide range of different detector types employed in alpha-particle spectrometry today includes semiconductor detectors, liquid scintillation counters, ionization chambers, proportional counters, and magnetic spectrometers. The most commonly used are silicon surface barrier (SSB) detectors and passivated ion-implanted silicon (PIPS) detectors. Low-level liquid scintillation counting, using pulse-shape discrimination to separate alpha- from beta-particle–induced events and anticoincidence shielding to reduce background, is also employed. The disadvantages are inferior energy resolution and sensitivity, compared to PIPS- or SSB-based systems, and problems associated with chemiluminescence [30].

2.7.9.2 Electrons

The interaction of electrons with matter is similar to the interaction of heavy particles with the following differences:

1. Nuclear collisions are not part of the interaction because of the very light electron mass.
2. At energies higher than a few MeV, radiative processes (bremsstrahlung) must be considered in addition to the inelastic electron collision.

Again, because of their light mass, electrons are so intensely scattered that their trajectory in the material is a jagged line; therefore, the concept of range is introduced. This is done by means of absorption experiments, which permit definition of the absorber thickness resulting in zero electron transmission at a given energy [1].

2.7.9.3 Gamma and X-Rays

A radioactive element may emit gamma rays (photons) if the nucleus remaining after alpha or beta decay is in an excited state. The three most important absorption processes are the photoelectric effect, the Compton effect, and the pair-production effect. The corpuscular description of electromagnetic radiation is the most appropriate for these effects, as one photon in a well-collimated beam of N_0 photons disappears at each interaction [1]. The attenuation of the photon beam can be described by a simple exponential law:

$$N = N_0 \exp(-\mu x) \qquad (2.17)$$

where N is the remaining photons in the beam after traversing distance x and the absorption coefficient μ is the sum of three terms due to the three processes mentioned earlier.

In the photoelectric interaction, the photon ejects a bound electron from an atom. All of the photon energy, $h\nu$, is given to the atom, which ejects electrons with energy $h\nu - E_1$, where E_1 is the binding energy of the electron. The excited atom then releases energy E_1 by decaying to its ground state. In this process, the atom releases one or more photons (and possibly an electron, called an Auger electron). The cross section of the photoelectric effect increases rapidly with the atomic number Z and decreases with increasing energy.

The Compton effect is essentially an elastic collision between a photon and an electron. During this interaction, the photon gives a fraction of its energy to the electrons, and its frequency ν is therefore decreased. The cross section for this effect decreases with increasing energy, but the decrease is less rapid than for the photoelectric effect [1].

In the pair-production effect, a high-energy photon near a nucleus gives up its energy to produce an electron-positron pair. The photon energy goes into the rest-mass energy and the kinetic energy of the electron-positron pair. The minimum energy necessary for this effect is set by elementary relativistic considerations at the value of 1.022 MeV, an amount equivalent to two electron rest masses. The cross section P for pair production increases with energy. Up to energies of 10 MeV, the P/Z ratio remains constant with energy [67]. At higher energies, the cross section starts to decrease for increasing values of the atomic number.

Several techniques developed for monitoring the radiation vary with measuring parameters, dose range, desired accuracy, duration of measurements, environmental condition, and so forth. For a better understanding of the instrumentation behind various radiation detectors, one may refer to a book by Shani [54], which reviews the basics of many radiation dosimetry instrumentations and methods. However, due to variety of products and developers, the exact details on instrumentation can be provided only by manufacturer. The fabrication of semiconductor nuclear detectors involves a number of critical steps. These will typically include growth of high-resistivity material, slicing and polishing of the device volume, the application of metal contacts (including photolithographic processes), surface passivation to limit surface leakage currents, and finally packaging (including bonding to the external circuitry).

The best energy resolution in modern radiation detectors can be achieved in the semiconductor materials, where a comparatively large number of carriers for a given incident radiation event occurs. There are both natural and technological limits to how precisely the energy of a gamma-ray detection event can be registered by the detection system. The natural limit on the energy precision arises primarily from the statistical fluctuations associated with the charge production processes in the detector medium. The voltage integrity of the full-energy pulses can also be disturbed by electronic effects, such as noise, pulse pileup, and improper pole-zero settings. These electronic effects have become less important as technology has improved, but their potential effects on the resolution must be considered in the setup of a counting system.

The majority of interest has centered upon the use of silicon, germanium, CdTe, HgI_2, GaAs, and so forth [68]. The semiconducting materials absorb radiation, and the level of adsorption varies with the material and radiation energy. In these materials, the basic information carriers are electron-hole pairs, which form a measurable electric current. However, there are three major processes that lead to the absorption of radiation: at low energies (below 100 keV), the photoelectric effect dominates; at intermediate energies (0.1 to 1 MeV), the Compton effect dominates; and at high energies (above 1.02 MeV), pair production dominates. The ionization energy is the average energy expended by the primary charged particle that produces one electron-hole pair [48]. Several configurations of silicon diodes are currently produced. Among them are diffused junction diodes, surface barrier diodes, ion implanted detectors, and epitaxial layer detectors. An alphabetical list of a number of companies (with their Web addresses) that develop, sell, or manufacture various radiation detection instruments can be found in the Appendix.

Si and Ge have long been used as semiconductor materials for radiation detection. However, due to their low band-gap energies, they cannot be used at room temperature. The band-gap energy of CdTe is wide ($E_g = 1.5$ eV); therefore, it is suitable for being used at room temperature. Materials like CdO, TeO, and TeO_2 are stable under normal conditions and are also suitable materials for radiation detectors [69]. An intensive research now is underway into new wide band-gap materials to enable yet more efficient and cost-effective radiation detection. Discussion on these materials is continued in Section 2.7.11.4.

2.7.10 Response of Diodes to External Exposure

P-n junctions are an integral part of several optoelectronic devices, which include photodiodes, solar cells, light emitting diodes (LEDs), semiconductor lasers, and radiation sensors. Photodiodes and crystalline solar cells are the p-n diodes, which, when exposed to light, yield a photocurrent in addition to the diode current, so that the total diode current is given by:

$$I = I_s \left(e^{V_a/V_t} - 1 \right) - I_{ph} \qquad (2.18)$$

where the additional photocurrent I_{ph} is due to photogeneration of electrons and holes. These electrons and holes are pulled into the region, where they are majority carriers by the electric field in the depletion region [70]. The photogenerated carriers cause a photocurrent, which opposes the diode current under forward bias. Therefore, the diode can be used as a photodetector (under reverse or zero bias voltage), as the measured photocurrent is proportional to the incident light intensity. The diode can also be used as a solar cell (using a forward bias) to generate electrical power. The dark current is the current through the diode in the absence of light. This current is due to the ideal diode current, the generation-recombination of carriers in the depletion region, and any surface leakage that occurs in the diode.

The trapping levels in the energy gap have a marked effect in speeding up the rate of recombination of excess minority current carriers [32]. When external excitation, such as light or radiation, interacts with the semiconductor material, it raises a number of electrons from the valence band to the conduction band, creating an equal number of free holes and electrons. Since the temperature is not increased, the energy width of the Fermi occupation probability is the same as it was before the external excitation. But now this is quasi-Fermi probability function for electrons and holes, respectively. A greater intensity of light or radiation will cause the two quasi-Fermi levels to separate farther. The greater external excitation will increase the percentage occupancy of the states of both bands and require adjustments of both quasi-Fermi levels. The region of the band gap between the two quasi-Fermi levels is no longer in thermal equilibrium with the enhanced free-carrier densities. Any impurity energy states laying in this region can act as trapping centers for the recombination of holes and electrons. External excitation generates excess densities of minority electrons in the p-type region and minority holes in the n-type region. These excited densities are able to influence the population of any deep trapping levels throughout the excited volume near the p-n junction, including the space charge region itself. On removal of the perturbing influence, the return from this excited state toward equilibrium occurs by three principal process [32]:

- The free excess densities of electrons Δn and of holes Δp decay by recombination with their characteristic lifetimes τ_e and τ_h, respectively, which are of the order of μs to ms depending on the quality of the material in question.
- If current is allowed to flow in the system, sweeping out may take place in times of the order of transit times τ_t in the n- and p-regions, which may be much shorter than the lifetimes τ_e and τ_h.
- Carriers trapped in the deep levels must normally be excited back into the relevant free band before being able to recombine—this process requires thermal excitation and may take many seconds, or more, depending on the temperature and the depth of the trap level.

A number of various semiconductor diode configurations can be used for radiation detection, depending on available technology and applications. These include diffused junctions, surface barriers, ion-implanted detectors, totally depleted detectors, and epitaxially fabricated diodes [1]. For penetrating radiations, such as gamma rays or neutrons, the damage is generally distributed throughout the detector, and the direction of incidence of the radiation has little effect. For electrons or charged particles, however, the orientation with respect to the detector is important. In general, diffused junction detectors are somewhat less susceptible to radiation damage effects than surface barriers. Also, fully depleted detectors are less sensitive than partially depleted devices. In turn, lithium drifted germanium detectors are more sensitive to radiation damage due to comparatively low average fields.

With the development of new technologies, materials, and fabrication methods, novel semiconductor diodes are being developed to deliver even better performance for radiation detection and measurements. The radiation hardness of particle detectors made with high-resistivity Czochralski silicon has been investigated with several irradiation campaigns, including low- and high-energy protons, neutrons, γ-rays, lithium ions, and electrons [71]. The performance of Czochralski silicon (Cz-Si), standard Float Zone silicon (Fz-Si), and oxygenated Fz-Si as radiation-hardened devices was compared. Characteristically, Fz-Si has a low oxygen concentration because of the contactless, crucible-free crystal growth technique, which can be a drawback since oxygen has experimentally been found to improve the radiation hardness of silicon detectors [72, 73]. Recent developments in the crystal growth technology of Cz-Si (namely, the magnetic Czochralski method) have enabled the production of Cz-Si wafers with sufficiently high resistivity and with a well-controlled, high concentration of oxygen [71]. It is anticipated that such improvements can push well beyond the operational limits of present silicon detectors.

DT: Detection Technology company (http://www.deetee.com) designs, manufactures, and markets unique, high-performance silicon photodiodes, radiation detectors, related electronics, and detector modules. Detection Technology offers two series of high-performance diodes optimized for X-rays and alpha, beta, and gamma radiation, namely the XRA and XRB series. Detection Technology's ultraclean processing method provides excellent performance characteristics and high reliability. The use of a unique guard-ring concept offers impressive performance improvements. The XRA series high-resistivity diodes are optimized for ionizing radiation detection and measurement applications. These diodes are characterized by an excellent response to weakly penetrating radiation achieved by the use of high-purity starting material and special ultraclean manufacturing process. Ultrathin radiation entrance window provides optimized response to low-energy X-rays and heavy charged particles. The XRB series high-resistivity diodes are optimized for ionizing radiation detection.

Excellent performance is based on the use of a high-resistivity silicon starting material and a high-purity manufacturing process. High radiation tolerance and depressed light sensitivity are achieved by the use of a specially developed radiation entrance window processing.

2.7.10.1 RADOS RAD-60 Personal Alarm Dosimeter

The RADOS RAD-60 Personal Alarm Dosimeter (Figure 2.21) from S. E. International (http://www.seintl.com) is a precise and reliable instrument for ensuring the personal safety of the user. It uses energy-compensated Si diode for gamma- and X-ray detection in the energy range of 60 keV–3 MeV, dose range 1 μSv–9.99 Sv, and dose rate 5 μSv/h–3 Sv/h. The RAD-60 can be switched into system mode, for the purpose of tracking personnel dose records and generating compliance reports. The design includes state-of-the-art technology with built-in memory for retrieving dose, even during power-down. It eliminates outside interference from shock and RF. The RAD-60 is easily programmed by the user, has a digital display, and operates with a single AAA alkaline battery (http://www.seintl.com/english/rados.htm).

Figure 2.21 RADOS RAD-60 Personal Alarm Dosimeter. (Courtesy of S.E. International, Inc.)

2.7.10.2 Siemens EPD-2 Electronic Personnel Dosimeter

The Siemens EPD-2 electronic personnel dosimeter is a direct reading electronic dosimeter that is sensitive to x- and γ-radiation and to β-particles. The applications of this dosimeter include support of emergency response operations. The EPD-2 detects radiation in the energy range from 15 keV to 10 MeV by use of multiple diode detectors, giving direct readout of dose equivalents $H_p(10)$ (deep/whole body) and $H_p(0.07)$ (shallow/skin) in units of Sv and rem. The dosimeter includes a number of audible and visual alarms for dose, dose rate, and other parameters that are user configurable via an infrared interface.

2.7.11 Other Radiation-Sensitive Materials of Interest

2.7.11.1 Polymers and Polymer Gels

Complex radiotherapy delivery techniques require dosimeters that are able to measure complex three-dimensional dose distributions accurately and with good spatial resolution. Polymer gel is an emerging new dosimeter being applied to these challenges. These dosimeters consist of a gel matrix within which a solution of acrylic molecules is suspended. These molecules polymerize upon exposure to radiation, with the degree of polymerization being proportional to absorbed dose. The polymer distribution can be measured in two or three dimensions using magnetic resonance imaging (MRI) or optical tomography. After calibration, the images can be converted into radiation dose distributions. Manufacture of the gel is reported to be reproducible, with accuracy of within 3±5% of measured dose in the range 0–10 Gy [74]. The development of radiation-sensitive gels has a long history, starting in 1950 with gels containing Folin's phenol, which change color upon irradiation [75]. Current interest in gel dosimetry technique is based on acrylic molecules embedded within a gel matrix, which polymerize upon irradiation, with the degree of polymerization being strongly related to the absorbed dose received by the gel. MRI or optical scanning methods can then image the localized polymerization with spatial resolution of ~1 mm. The relaxation rate of the polymerized region is linearly proportional to absorbed dose in a dose range applicable to the study of clinical radiation therapy (up to 15 Gy) [74].

Irradiation of polymers causes structural and chemical variations, which in turn lead to the variations in physical properties. Radiation-induced changes in optical properties of a number of polymers have been investigated for dosimetry applications, including cellulose triacetate [76], diacetylene Langmuir-Blodgett films [77], polyaniline nanofilms [78], and tetrafluoroethylene-per-fluoromethoxyethylene (PFA) and tetrafluoroethylene-hexa-fluoropropylene (FEP) thin films [79]. A comprehensive review of optical properties of 17 different polymers under irradiation can be found in [80].

2.7.11.2 Diamond Dosimeters

Diamonds change their resistance upon radiation exposure. When applying a bias voltage, the resulting current is proportional to the dose rate of radiation. Commercially available diamond dosimeters are designed to measure relative dose distributions in high-energy photon and electron beams. The dosimeter is based on a natural diamond crystal sealed in polystyrene housing with a bias applied through thin golden contacts. Diamonds have a small sensitive volume, of the order of a few mm^3, which allows the measurement of dose distributions with an excellent spatial resolution. Diamond dosimeters are tissue-equivalent and require nearly no energy correction. Due to their flat energy response, small physical size, and negligible directional dependence, diamonds are well suited for use in high-dose gradient regions (e.g., for stereotactic radiosurgery). Stereotactic radiosurgery can be used in patients who have failed standard radiation techniques or in patients who have already received the maximum radiation dose permissible or whole brain radiation.

2.7.11.3 Alanine

Alanine is one of the amino acids, and when pressed in the form of rods or pellets with an inert binding material, it is typically used for high-dose dosimetry, at a level of about 10 Gy or more with sufficient precision for radiotherapy dosimetry. The radiation interaction results in the formation of alanine radicals, the concentration of which can be measured using an electron spin resonance spectrometer. The intensity is measured as the peak-to-peak height of the central line in the spectrum. The readout is nondestructive. Although alanine is tissue-equivalent and requires no energy correction within the quality range of typical therapeutic beams, its response depends on storage and environmental conditions during irradiation, such as temperature and humidity.

2.7.11.4 CZT Materials

Among the semiconductor materials, cadmium telluride (CdTe) and $Cd_{1-x}Zn_xTe$ (CZT) have attracted the most attention as room-temperature X-ray and γ-ray detectors, due to their unique combination of high atomic number (good stopping power), good mobility lifetime product (good transport properties of the free charge carriers), and large band gap [81]. CdTe and ZnTe have the cubic zinc sulfide, or zincblende, structure. This structure is sometimes described as a pair of interpenetrating face centered cubic (fcc) sublattices, offset from each other by one quarter of a unit cell body diagonal, with the Cd or Zn nuclei occupying one sublattice and the Te nuclei occupying the other. Interest in CZT arises from the ability to create nearly perfect lattice matching with such materials as HgCdTe, HgZnTe, and HgZnSeTe by adjustment of the Zn composition (x). In addition to lattice matching, a high degree of structural perfection is necessary. For room temperature nuclear detector applications, very stringent electrical requirements must be met, as well as a high degree of structural perfection. Three major growth methods at different stages of industrial development are now available to produce spectrometers with good resolution over a wide energy range: CdTe:Cl THM (Traveling Heater Method), CdZnTe HPBM (High Pressure Bridgman Method), and CdZnTe:In HBM (Horizontal Bridgman Method). The primary goal on which every growth technique is focused is the production of defect-free, high-resistivity material, in sufficient quantities to ensure high yield and low cost. Material resistivity is typically controlled by intrinsic and extrinsic dopants, which pin the Fermi level near midgap. The research continues on alternative growth methods, such as physical vapor deposition (PVD) and other melt-growth methods [82].

The structural designs of detectors affect their spectroscopic characteristics and include hemispherical detectors, coplanar strip-electrode detectors, and monolithic, two-dimensional segmented electrode arrays with pad sizes smaller than their thickness. Simple planar detectors or coplanar grids are generally employed for large-volume single-element detectors. The former is simpler to implement in terms of the associated electronics, but the other one operates as an electron-only device and provides higher spectral resolution, particularly for higher energy gamma rays (>200 keV). Pixellated detectors are employed in imaging systems, where position information is contained in the signals from individual pixels. In each case, lithographic techniques are employed to define the geometry of the metal electrodes, which are applied to the surface of the semiconductor material. Type and structure of electrodes, namely, metal-semiconductor contacts, affect radiation sensors parameters; therefore, they are discussed in more detail in Section 3.5 of Chapter 3.

In medical applications, CdTe and CZT semiconductor materials are widely employed for flat-panel digital detectors used in digital radiography, computerized tomography, fluoroscopy, radionucleide imaging, and nuclear medicine. These systems are comprised of large-area pixel arrays, which use matrix addressing to read out charges resulting from X-ray absorption in the detector medium. Emerging applications of CZT materials in medicine include chest X-ray imaging, high-resolution dental digital radiography systems, mammography, bone densitometry, and γ-cameras [83]. One may refer to a comprehensive review of the CZT material properties and fabrication techniques related to room temperature nuclear detectors [82].

Multipurpose Radiation Monitor PM1401K The PM1401K http://www.polimaster. com) is a novel multipurpose pocket-type monitor, which performs all procedures of the radiation control process and transmits the results of measurements to computer networks through radio channels for further processing, analysis, and formation of the correspondent databases. PM1401K detects α, β, γ, and neutron radiation (see Figure 2.22). It is also a spectrometer (1,024 channel)

Figure 2.22 Multipurpose Radiation Monitor PM1401K. (Courtesy of Polimaster, Inc.)

and may work as a real-time radioisotope identifier. For the detection of gamma radiation, it uses a CsI(Tl) sensor in the energy range of 0.06–3.0MeV and a GM counter in the energy range of 0.01–515 MeV. For the detection of neutrons, it uses a He-3 neutron counter with a moderator. For measuring α and β radiation, a special GM counter is employed, with the minimum detectable α-flux density of 2 min^{-1} cm^{-2} and β-flux density measurement range from 6 $min^{-1}cm^{-2}$ to 10^5 $min^{-1}cm^{-2}$.

Polismart The Polismart Integrated Radiation Control System was created by Polimaster (http://www.polimaster.com) and consists of unlimited number of independent remote personal radiation detectors that can identify nuclear and radioactive materials and operate in wireless communication with the expert decision support system in the real time. The portable multipurpose instruments that are included into POLISMART system consist of two main modules: an independent intellectual radiation detector unit and a unit for processing, displaying, and transmitting the information to the decision support system. The instrument may include detectors of different sensitivity for measurement, search, and identification of gamma and neutron radiation sources. The different models of smart phones or pocket PCs can be used as the unit for processing, displaying, and transmitting the information.

The PM1802 (Figure 2.23) is the personal radiation detector-identifier with two independent gamma and neutron channels. The PM1802 can be used both as an independent radiation control instrument and along with the fixed-installed radiation monitors. As it is equipped with two separate scintillation

Figure 2.23 Gamma-neutron PM1802. (Courtesy of Polimaster, Inc.)

detectors based on CsI(Tl) and LiI(Eu), the PM1802 is capable of operating as a two-channel alarming rate meter, ensuring fast location and measurement of gamma and neutron radiation sources, and performing fast identification of the gamma and neutron sources on their gamma spectra in the field conditions. The PM1802 has an energy range of 0.03–33.0 MeV for gamma radiation and thermal to 14 MeV for neutron detection. Detailed technical information for a number of Polimaster developments can be found at http://www.polimaster.com.

2.7.11.5 Optical Fiber Sensors

While the technology for detection and identification of radioactive material using gamma-ray spectrometers based on scintillators (see Section 2.7.4) or semiconductor crystals (see Section 2.7.9) is relatively well-established, a variety of situations may require more simple and inexpensive radiation sensors for widespread monitoring that can be remotely interrogated at regular periods with little or no human intervention and are easy to install, operate, and maintain. Radiation sensors based on thin and thick films of metal oxides and metal-substituted phthalocyanines are considered an emerging technology for cost-effective dosimetry, and they are discussed in Chapters 4 and 5.

Optical fiber dosimeters to remotely measure absorbed dose due to exposure to ionizing radiation are of current interest. Such devices may operate, for example, as trigger detectors, operating at low powers, for subsequent intervention with more detailed analytical technologies. Utilization of fiber-optic technology provides the potential for making highly spatially resolved real-time measurements of the absorbed dose. OSL (see Section 2.7.7.2) allows the use of optical fibers to remotely interrogate the radiation sensor. The fiber can guide the stimulation light to the radiation sensor, installed in a probe at a remote location, and can carry the OSL signal emitted by the sensor back to a portable or central reader. The probes themselves are inexpensive, consisting simply of a radiation-sensitive crystal coupled to the optical fiber. Multiple probes, installed at different locations covering a large area or structure, can in principle be connected to a single multiplexing reader configured for periodic readouts, creating an integrated sensor network that monitors the radiation level at these locations. Each probe can also be independently read using a portable reader. One application that could benefit from such a system is the monitoring of radioactive plumes from radioactive waste storage sites, including continuous verification that nuclear waste sites are not leaking. Inexpensive subsurface remote probes connected to a central station or to a portable reader could detect and quantify radioactive plumes during remediation efforts, verify the absence of radioactive materials, and provide assurance that radioactive material is not leaking from the site [84]. Besides application for the detection of radiological materials and contaminants, the system can also be used as a sensitive remote dosimeter for areas of difficult access or hazardous locations, such as the ground water around nuclear storage facilities or high-radiation-level areas in and around nuclear reactors. Additional potential applications for remote fiber-optic dosimeters include on-site determination of the gamma component of the natural dose rate in soils for general environmental monitoring applications, in vivo patient dose verification during cancer radiotherapy treatment, and remote radiation detection for security purposes and for the prevention of terrorist attacks [84, 85]. Fiber-optic coupled scintillation detectors provide an instantaneous measurement of the dose rate, and the total dose can be readily calculated by integration of the dose rate data over time.

References

[1] Knoll, G. F., *Radiation Detection and Measurement*, New York: John Wiley & Sons, 1979.

[2] Holmes-Siedle, A. G., and L. Adams, *Handbook of Radiation Effects*, 2nd ed., New York: Oxford University Press, 2002.

[3] ICRU Report 49, *Stopping Powers and Ranges for Protons and Alpha Particles,* International Commission on Radiation Units and Measurements, 1993.

[4] Bethe, H., "Theory of the Passage of Rapid Corpuscular Rays Through Matter," *Annalen der Physik*, Vol. 5, 1930, pp. 325–400.

[5] ICRU Report 37, *Stopping Powers for Electrons and Positrons*, International Commission on Radiation Units and Measurements, 1984.

[6] ICRU Report 16, *Linear Energy Transfer*, International Commission on Radiation Units and Measurements, 1970.

[7] Dörschel, B., V. Schuricht, and J. Steuer, *The Physics of Radiation Protection*, Ashford, Kent, U.K.: Nuclear Technology, 1996.

[8] NCRP Report No. 82, *SI Units in Radiation Protection and Measurements*, National Council on Radiation Protection and Measurements, 1985.

[9] NCRP, "Evaluation of the Linear-Nonthreshold Dose-Response Model for Ionizing Radiation," *NCRP*, Vol. 136, 2001.

[10] ICRP, "ICRP Publication 60: 1990 Recommendations of the International Commission on Radiological Protection 60," *Annals of the ICRP*, 1991.

[11] Makhijani, A., and B. Franke, "Worker Radiation Doses Deeply Flawed," *Science for Democratic Action, IEER Newsletter*, Vol. 6, 1997.

[12] ICRP, "ICRP Publication 92: Relative Biological Effectiveness (RBE), Quality Factor (Q), and Radiation Weighting Factor (WR)," *Elsevier: International Commission on Radiological Protection*, 2004.

[13] NCRP Report No. 116, *Limitation of Exposure to Ionizing Radiation*, National Council on Radiation Protection and Measurements, 1993.

[14] Miller, A., P. Sharpe, and R. Chu, *Dosimetry for Industrial Radiation Processing*, International Commission on Radiation Units and Measurements, 2000.

[15] International Organization for Standardization and Case Postal 56, C. G. 2. S., "Sterilization of Health Care Products," *Requirements for Validation and Routine Control Radiation Sterilization*, Vol. ISO-11137, 1995.

[16] European Committee for Standardization, "Sterilization of Medical Devices," *Validation and Routine Control of Sterilization by Irradiation*, Vol. EN 552, 1994.

[17] Machi, S., "Prospects for the Application of Radiation Processing and the Activities of the IAEA," *Radiation Physics and Chemistry*, Vol. 52, No. 1–6, 1998, pp. 591–597.

[18] Descamps, T., "The Practical Experience of a Total Conversion to High Energy Electron Beam Processing," *Radiation Physics and Chemistry*, Vol. 46, No. 4–6, 1995, pp. 439–442.

[19] Clough, R. L., "High-Energy Radiation and Polymers: A Review of Commercial Processes and Emerging Applications," *Nuclear Instruments and Methods in Physics Research Section B: Beam Interactions with Materials and Atoms*, Vol. 185, No. 1–4, 2001, pp. 8–33.

[20] Ross, R. T., and D. Engeljohn, "Food Irradiation in the United States: Irradiation as a Phytosanitary Treatment for Fresh Fruits and Vegetables and for the Control of Microorganisms in Meat and Poultry," *Radiation Physics and Chemistry*, Vol. 57, 2000, pp. 211–214.

[21] Clough, R. L., and S. W. Shalaby, *Irradiation of Polymers: Fundamentals and Technological Applications*, Washington, D.C.: American Chemical Society, 1996.

[22] Singh, A., and J. Silverman, *Radiation Processing of Polymers*, Munich: Hanser, 1992.

[23] Kurtz, S. R., and C. Arnold, "Photocarrier Transport and Trapping Processes in Doped Polyethylene Terephthalate Films," *Journal of Applied Physics,* Vol. 57, 1985, pp. 2532–2537.

[24] Watson, S. J., et al., *Ionising Radiation Exposure of the UK Population: 2005 Review,* Health Protection Agency Report, Vol. HPA-PRD-001, 2005.

[25] Rosenfeld, A. B., et al., "In Vivo Dosimetry and Seed Localization in Prostate Brachytherapy with Permanent Implants," *IEEE Trans. on Nuclear Science,* Vol. 51, No. 6 I, 2004, pp. 3013–3018.

[26] Consorti, R., et al., "In Vivo Dosimetry with MOSFETs: Dosimetric Characterization and First Clinical Results in Intraoperative Radiotherapy," *International Journal of Radiation Oncology Biology Physics,* Vol. 63, No. 3, 2005, pp. 952–960.

[27] Price, R. A., et al., "Development of a RadFET Linear Array for Intracavitary In Vivo Dosimetry During External Beam Radiotherapy and Brachytherapy," *IEEE Trans. on Nuclear Science,* Vol. 51, No. 4 I, 2004, pp. 1420–1426.

[28] O'Dea, J., *Exposure: Living with Radiation in Ireland,* Dublin: Irish Reporter Publications, 1997.

[29] "Lost and Stolen Nuclear Materials in the United States," http://www.ki4u.com/lost.htm, 2001.

[30] Warner, F., and R. M. Harrison, *Radioecology After Chernobyl,* SCOPE 50: Scientific Committee On Problems of the Environment, 1993.

[31] Lieberman, B., "Contaminated Machine Held in Former Nuclear Worker's Home," *Harrisburg Patriot News,* October 27, 1996.

[32] Hunter, L. P., *Handbook of Semiconductor Electronics,* New York: McGraw-Hill, 1970.

[33] International Electrotechnical Commission, "Radiation Protection Instrumentation Measurement of Personal Dose Equivalent Hp(10) and Hp(0.07) for X, Gamma and Beta Radiations—Direct Reading Personal Dose Equivalent and/or Dose Equivalent Rate Dosemeters," *International Standard IEC 61526,* July 1998.

[34] Barthe, J., "Electronic Dosimeters Based on Solid State Detectors," *Nuclear Instruments and Methods in Physics Research Section B: Beam Interactions with Materials and Atoms,* Vol. 184, No. 1–2, 2001, pp. 158–189.

[35] Nagarkar, V. V., et al., "Structured CsI(Tl) Scintillators for X-Ray Imaging Applications," *IEEE Trans. on Nuclear Science,* Vol. 45, No. 3, 1998, pp. 492–496.

[36] Harris, E. J., et al., "CCD-Based γ-Ray Dosimeter," *Nuclear Instruments and Methods in Physics Research Section A: Accelerators, Spectrometers, Detectors and Associated Equipment,* Vol. 458, No. 1–2, 2001, pp. 227–232.

[37] McDonald, J. C., "Calibration Measurements and Standards for Radiation Protection Dosimetry," *Radiation Protection Dosimetry,* Vol. 109, No. 4, 2004, pp. 317–321.

[38] ISO 17025, *General Requirements for the Competence of Testing and Calibration Laboratories,* Geneva: International Organization for Standardization, 1999.

[39] ISO 4037-1, *X and Gamma Reference Radiations for Calibrating Dosemeters and Doserate Meters and for Determining Their Response as a Function of Photon Energy—Part 1: Radiation Characteristics and Production Methods,* Geneva: International Organization for Standardization, 1996.

[40] ISO 4037-2, *X and Gamma Reference Radiations for Calibrating Dosemeters and Doserate Meters and for Determining Their Response as a Function of Photon Energy—Part 2: Dosimetry for Radiation Protection over the Energy Ranges from 8 keV to 1.3 MeV and 4 MeV to 9 MeV,* Geneva: International Organization for Standardization, 1997.

[41] ISO 4037-3, *X and Gamma Reference Radiations for Calibrating Dosemeters and Doserate Meters and for Determining Their Response as a Function of Photon Energy—Part 3: Calibration of Area and Personal Dosemeters and the Measurement of Their Response as a Function of Energy and Angle of Incidence,* Geneva: International Organization for Standardization, 1999.

[42] ISO 4037-4, *X and Gamma Reference Radiations for Calibrating Dosemeters and Doserate Meters and for Determining Their Response as a Function of Photon Energy—Part 4: Calibration of Area and Personal Dosemeters in Low Energy X Reference Radiation Fields,* Geneva: International Organization for Standardization, 2004.

[43] ISO 6980, *Reference Beta Radiations for Calibrating Dosemeters and Doserate Meters and for Determining Their Response as a Function of Beta-Radiation Energy,* Geneva: International Organization for Standardization, 1996.

[44] ISO 6980-2, *Nuclear Energy—Reference Beta-Particle Radiation—Part 2: Calibration Fundamentals Related to Basic Quantities Characterizing the Radiation Field,* Geneva: International Organization for Standardization, 2004.

[45] ISO 8529-1, *Reference Neutron Radiations—Part 1: Characteristics and Methods of Production,* Geneva: International Organization for Standardization, 2001.

[46] ISO 8529-2, *Reference Neutron Radiations—Part 2: Calibration Fundamentals of Radiation Protection Devices Related to the Basic Quantities Characterizing the Radiation Field,* Geneva: International Organization for Standardization, 2000.

[47] ISO 8529-3, *Reference Neutron Radiations—Part 3: Calibration of Area and Personal Dosimeters and Determination of Response as a Function of Energy and Angle of Incidence,* Geneva: International Organization for Standardization, 1998.

[48] Fraden, J., *Handbook of Modern Sensors: Physics, Designs, and Applications,* New York: Springer, 1996.

[49] Gektin, A. V., "Scintillators and Storage Phosphors Based on ABX3 Crystals," *Journal of Luminescence,* Vol. 87–89, 2000, pp. 1283–1285.

[50] Alcon, E. P. Q., R. T. Lopes, and C. E. V. de Almeida, "EPR Study of Radiation Stability of Organic Plastic Scintillator for Cardiovascular Brachytherapy Sr90-Y90 Beta Dosimetry," *Applied Radiation and Isotopes,* Vol. 62, No. 2, 2005, pp. 301–306.

[51] Li, Z., et al., "Properties of Plastic Scintillators After Irradiation," *Nuclear Instruments and Methods in Physics Research Section A: Accelerators, Spectrometers, Detectors and Associated Equipment,* Vol. 552, No. 3, 2005, pp. 449–455.

[52] Becks, K.-H., et al., "A Multi Channel Dosimeter Based on Scintillating Fibers for Medical Applications," *Nuclear Instruments and Methods in Physics Research Section A: Accelerators, Spectrometers, Detectors, and Associated Equipment,* Vol. 454, No. 1, 2000, pp. 147–151.

[53] Jayson, G. G., B. J. Parsona, and A. J. Swallow, "The Mechanism of the Fricke Dosimeter," *International Journal for Radiation Physics and Chemistry,* Vol. 7, No. 2–3, 1975, pp. 363–370.

[54] Shani, G., *Radiation Dosimetry Instrumentation and Methods*, Boca Raton, FL: Interpharm/CRC, 1991.

[55] Martinez-Davalos, A., et al., "Radiochromic Dye Film Studies for Brachytherapy Applications," *Radiation Protection Dosimetry*, Vol. 101, No. 1–4, 2002, pp. 489–492.

[56] Fowler, W. B., *Physics of Color Centres*, New York: Academic, 1968.

[57] Gamboa-deBuen, I., et al., "Supralinear Response and Efficiency of LiF:Mg,Ti to 0.7, 1.5 and 3 MeV Protons," *Nuclear Instruments and Methods in Physics Research Section B: Beam Interactions with Materials and Atoms*, Vol. 183, No. 3–4, 2001, pp. 487–496.

[58] Horowitz, Y. S., O. Avila, and M. Rodriguez-Villafuerte, "Theory of Heavy Charged Particle Response (Efficiency and Supralinearity) in TL Materials," *Nuclear Instruments and Methods in Physics Research Section B: Beam Interactions with Materials and Atoms*, Vol. 184, No. 1–2, 2001, pp. 85–112.

[59] Horowitz, Y. S., "Theory of Thermoluminescence Gamma Dose Response: The Unified Interaction Model," *Nuclear Instruments and Methods in Physics Research Section B: Beam Interactions with Materials and Atoms*, Vol. 184, No. 1–2, 2001, pp. 68–84.

[60] McKeever, S. W. S., and R. Chen, "Luminescence Models, Radiation Measurements," Vol. 27, No. 5–6, 1997, pp. 625–661.

[61] Sunta, C. M., et al., "Thermally Stimulated Luminescence and Conductivity—Theoretical Models and Their Applicability to Experimental Results," *Radiation Effects and Defects in Solids*, Vol. 146, No. 1–4, Pt. 1, 1998, pp. 261–276.

[62] Chernov, V., B. Rogalev, and M. Barboza-Flores, "Dose Rate Effect on the Yield of Radiation Induced Response with Thermal Fading," *Radiation Measurements*, Vol. 39, No. 3, 2005, pp. 329–335.

[63] Shachar, B. B., G. L. Catchen, and J. M. Hoffmann, "Characteristics of the Panasonic UD-802 Phosphors," *Radiation Protection Dosimetry*, Vol. 27, No. 2, 1989, pp. 121–124.

[64] Lee, J. I., et al., "Dosimetric Properties of the Newly Developed KLT-300 (LiF:Mg,Cu,Na,Si) TL Detector," *Radiation Measurements*, Vol. 38, No. 4–6, 2004, pp. 439–442.

[65] Holmes-Siedle, A. G., and L. Adams, *Handbook of Radiation Effects*, New York: Oxford University Press, 1993.

[66] Oldham, T. R., *Ionizing Radiation Effects in MOS Oxides*, Hackensack, NJ: World Scientific Publishing, 2000.

[67] Goulding, F. S., and R. H. Pehl, "Nuclear Spectroscopy and Reactions," in *Semiconductor Detectors*, J. Cerny, (ed.), New York: Academic Press, 1974.

[68] Gardner, J. W., *Microsensors: Principles and Applications*, New York: John Wiley & Sons, 1994.

[69] Baba, S., et al., "Recent Development of Radiation Measurement Instrument for Industrial and Medical Applications," *Nuclear Instruments and Methods in Physics Research Section A: Accelerators, Spectrometers, Detectors and Associated Equipment*, Vol. 458, 2001, pp. 262–268.

[70] Kasap, S. O., *Principles of Electronic Materials and Devices*, New York: McGraw-Hill, Burr Ridge, 2002.

[71] Harkonen, J., et al., "Particle Detectors Made of High-Resistivity Czochralski Silicon," *Nuclear Instruments and Methods in Physics Research Section A: Accelerators, Spectrometers, Detectors and Associated Equipment*, Vol. 541, No. 1–2, 2005, pp. 202–207.

[72] Kuhnke, M., E. Fretwurst, and G. Lindstroem, "Defect Generation in Crystalline Silicon Irradiated with High Energy Particles," *Nuclear Instruments and Methods in Physics Research Section B: Beam Interactions with Materials and Atoms*, Vol. 186, No. 14, 2002, pp. 144–151.

[73] Stahl, J., et al., "Radiation Hardness of Silicon—A Challenge for Defect Engineering," *Physica B: Condensed Matter*, Vol. 340–342, 2003, pp. 705–709.

[74] McJury, M., et al., "Radiation Dosimetry Using Polymer Gels: Methods and Applications," *The British Journal of Radiology*, Vol. 73, No. 873, 2000, pp. 919–929.

[75] Day, M. J., and G. Stein, "Chemical Effects of Ionising Radiation in Some Gels," *Nature*, Vol. 166, No. 141, 1950, p. 7.

[76] El Fiki, S. A., et al., "Investigation of the Effect of Gamma Rays on Optical Properties of Polymers," *Radiation Physics and Chemistry*, Vol. 47, No. 5, 1996, pp. 761–764.

[77] Mod Ali, N., C. E. Tucker, and F. A. Smith, "Consideration of Radiation-Induced Polymerization of Diacetylene LB Films for Dosimetry," *Thin Solid Films*, Vol. 289, No. 12, 1996, pp. 267–271.

[78] Laranjeira, J. M. G., et al., "Polyaniline Nanofilms as a Sensing Device for Ionizing Radiation," *Physica E: Low-Dimensional Systems and Nanostructures*, Vol. 17, 2003, pp. 666–667.

[79] Parada, M. A., et al., "Effects of MeV Proton Bombardment in Thin Film PFA and FEP Polymers," *Surface and Coatings Technology*, Vol. 196, No. 1–3, 2005, pp. 378–382.

[80] Clough, R. L., et al., "Color Formation in Irradiated Polymers," *Radiation Physics and Chemistry*, Vol. 48, No. 5, 1996, pp. 583–594.

[81] Verger, L., et al., "Characterization of CdTe and CdZnTe Detectors for Gamma-Ray Imaging Applications," *Nuclear Instruments and Methods in Physics Research Section A: Accelerators, Spectrometers, Detectors and Associated Equipment*, Vol. 458, No. 1–2, 2001, pp. 297–309.

[82] Schlesinger, T. E., et al., "Cadmium Zinc Telluride and Its Use as a Nuclear Radiation Detector Material," *Materials Science and Engineering: R: Reports*, Vol. 32, No. 45, 2001, pp. 103–189.

[83] Scheiber, C., and G. C. Giakos, "Medical Applications of CdTe and CdZnTe Detectors," *Nuclear Instruments and Methods in Physics Research Section A: Accelerators, Spectrometers, Detectors and Associated Equipment*, Vol. 458, No. 1–2, 2001, pp. 12–25.

[84] Klein, D. M., et al., "An Optical Fiber Radiation Sensor for Remote Detection of Radiological Materials," *IEEE Sensors Journal*, Vol. 5, No. 4, 2005, pp. 581–588.

[85] Justus, B. L., et al., "Gated Fiber-Optic-Coupled Detector for In Vivo Real-Time Radiation Dosimetry," *Applied Optics*, Vol. 43, No. 8, 2004, pp. 1663–1668.

3

Effect of Radiation on Optical and Electrical Properties of Materials

3.1 Introduction

The primary interactions between energetic radiation, semiconductors, and inorganic insulators result in the loss of energy to their electrons, and this energy is ultimately converted to the form of electron-hole pairs. In this process, known as ionization, the valence band electrons in the solid are excited to the conduction band and are highly mobile, if an electric field is applied [1]. Thus, any solid conducts for a time at a higher level than normal. The production and subsequent trapping of the holes in oxide films cause serious alterations in the device's performance.

In organic materials under irradiation, a chain of reactions, in which oxygen and moisture from the environment may be incorporated, starts with rapid electronic phenomena. This is followed by the generation of reactive, short-lived intermediate compounds and a complex mixture of chemical products. The main result of ionization is the breaking of chemical bonds and the creation of new ones, which causes change in conductivity and leads to long-lived forms of physical breakdown. Changes in the electrical properties of metals under irradiation are considered mainly in nuclear reactor applications, where small conductivity alterations due to atom displacement might affect a carefully balanced microelectronic circuit [1].

Optical absorption analysis has widely proven to be an important and efficient tool in exploring and interpreting the various phenomena of electronic structures and processes in the materials subjected to radiation [2–4]. The considerable theoretical investigations on the optical behavior of thin films deal primarily with optical reflection, transmission, and adsorption properties, and their relation to the optical constants of films [2]. The importance of studying the optical properties of a material is offered by the ability of this technique to provide information regarding the fundamental gap, electronic transition, trapping levels, and localized states. In general, films are amorphous, and at most they are polycrystalline in nature. Over the last decades, advances have been made in understanding the problem of how the disorder in amorphous materials influences the band structure and hence the electrical and optical properties of the material. For semiconductors, the main characteristics of the energy distribution of electronic states density of the crystalline solids are the sharp structure in the valence and conduction bands, as well as the abrupt terminations at the valence band maximum and the conduction band minimum. The sharp edges in the density of states curves produce a well-defined forbidden energy gap. An amorphous solid is a material lacking any form of structural order. The amorphous state is to some extent unstable or meta-stable [5] and frequently exhibits a gradual or even rapid transition to an ordered crystalline condition. Despite the presence of high density disorder in amorphous materials, they depart only slightly from the ideal crystalline structure [5]. In other words, short-range order can be assumed due to the rigidity of the chemical bonds, and the fundamentals of crystal band structure still holds for amorphous solids [6]. Nominally amorphous films may differ in their electrical and other properties according to the manner of their preparation. In particular, the deposition rate in evaporated films is known to have a profound influence on the dielectric constant and the level of conductivity [7].

3.2 Optical Absorption

In an intrinsic system, the highest occupied continuous energy sublevels are called the valence band (E_V) and the lowest unoccupied continuous energy sublevels are known as the conduction band (E_C). Both bands are separated by a forbidden energy band gap, as shown in Figure 3.1. The energy band gap width is denoted E_g.

Optical absorption in solids occurs by various mechanisms, in all of which the photon energy is absorbed by either the lattice or by electrons, where the transferred energy is conserved. The lattice (or phonon) absorption gives information about the atomic vibrations involved. This absorption of radiation normally occurs in the infrared region of the spectrum. Higher energy parts could provide information associated with the interband electronic transition. In these processes, the electrons are excited from the filled band to an empty band by the photon absorption, and, as a consequence of this, a sharp increase in the absorption coefficient $\alpha(\nu)$ will result. The onset of this rapid change in $\alpha(\nu)$ is called

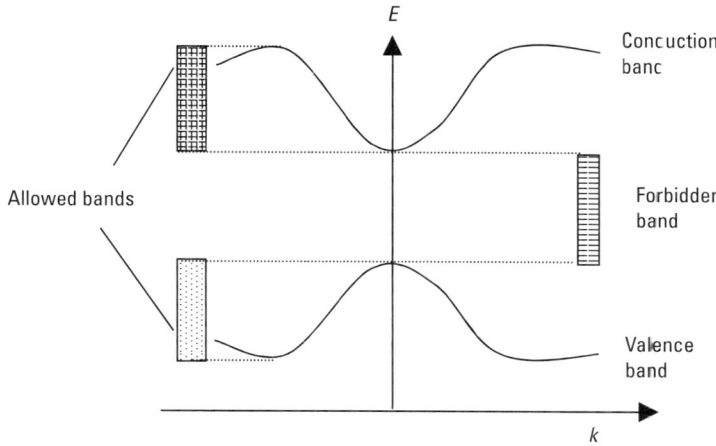

Figure 3.1 Representation of *E-k* energy band diagram.

the fundamental absorption edge and the corresponding energy is defined as the optical band gap $E_{opt.}$. Photons with a certain range of energy could be absorbed either due to internal transitions between the d-shell electrons or due to a transfer of an electron from the neighboring atom to the modifier ion (transition metal ion) and vice versa.

In real crystalline systems, impurity is an essential component of all materials, and the energy-wave vector (*E-k*) relation is more complex than the ideal intrinsic or compensated systems displayed in Figure 3.1. A more realistic view of the energy band structure is displayed in Figure 3.2, where the valence and conduction bands are nonsymmetric. In addition, the conduction band may have more than one minimum, and the valence band may also have more than one maximum.

There are two kinds of optical transitions at the fundamental absorption edge of crystalline and noncrystalline semiconductors: direct and indirect transitions, both of which involve the interaction of an electromagnetic wave with an electron in the valence band. The electron is then raised across the fundamental gap to the conduction band. For the direct optical transition from the valence band to the conductance band, it is essential that the wave vector for the electron be unchanged. In the case of an indirect transition, the interactions with the lattice vibrations (photons) take place; thus, the wave vector of the electron can change in the optical transition, and the momentum change will be taken or given up by photons. In other words, if the bottom of the conduction band lies at a different part of K-space from the top of the valence band, a direct transition from the top of the valence band to the bottom of the conduction band is forbidden. The direct and indirect optical transitions are determined mainly by type of the fundamental gap as follows:

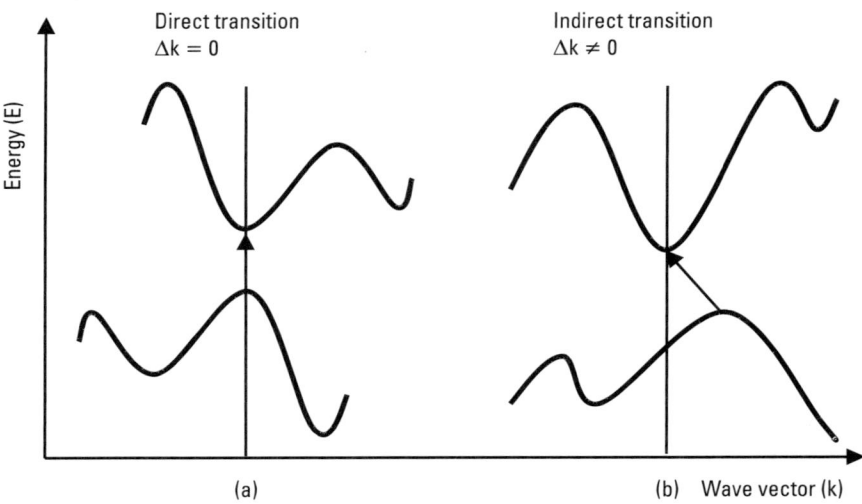

Figure 3.2 (a) Direct and (b) indirect transitions in fundamental gap for crystalline structure.

1. When the minimum of E_c and the maximum of E_v are aligned (i.e., $\Delta k = 0$), then the quantum number is said to be k-conserved, and the transition is known as direct transition [see Figure 3.2(a)]. Electrons of this transition move from the top of the valence band to the bottom of the conduction band, and vice versa, by absorbing or emitting a photon without involving any phonon interaction. The photon energy must be greater than or equal to $E_c - E_v$. The classical examples of direct transition semiconductors are GaAs, InP, InSb, and CdS.
2. In Figure 3.2(b), the bottom of the conduction band and the top of the valence band are misaligned and $\Delta k \neq 0$. The quantum number is not conserved, and the transition is said to be an indirect transition. The transport in this case involves simultaneously electron-photon and electron-phonon interactions when crossing the misaligned forbidden gap.

3.3 Amorphous Films

The structural order of a film is determined largely by the surface mobility of the adsorbed atoms. A highly disordered amorphous-like structure would be produced if the mobility is negligible so that the atoms condense at or near the point of impingement [2]. Since oxides have considerably lower atomic mobility than the metals at any temperature, no significant crystallization is expected to take place [8].

There are three main types of disorder. The first is described as a continuously connected long-range disorder, which can be observed for a single crystal. It involves no broken bonds, and no foreign impurities are present. The second type is connected with materials having a large density of broken bonds forming grain boundaries and leading to a discontinuous structure. The third type of disorder is due to foreign impurities, incorporated interstitially or by substitution. These impurities may have originated in the initial process of material manufacture or they may have been introduced by diffusion.

It would be appropriate to recall here the salient features of the theory of amorphous films. The general consensus of opinion is that the basic features of the band structure, such as the width of the forbidden gap, are determined primarily by the short-range order (i.e., by the relative dispositions of nearest neighbors in the solid) [7]. Since these dispositions are similar in amorphous and crystalline solids, the broad features are preserved on transition from crystalline to amorphous structure. The disappearance of medium- and long-range order does influence the detailed shape of the band structure. It causes a considerable blurring of the edges of the conduction and valence bands and gives rise to a distribution of deep localized levels (tail states) in the forbidden gap [7]. In a crystalline solid, a clear distinction exists between the conduction, valence band, and the forbidden gap. In both the conduction and the valence bands, charge carriers propagate freely, except for collisions due to thermal vibrations and other lattice imperfections [7]. In the forbidden gap, the energy levels that may exist due to imperfections are strictly localized, and an electron has to be excited from one of these levels into the conduction band before being able to move on. Special cases arise when the localized levels are spaced so closely that their wave functions overlap and give rise to the formation of an *impurity band*, leading to metallic properties with zero activation energy. Alternatively, the spacing may not be as close as is necessary for the formation of an impurity band, but it may be sufficient for phonon-assisted tunneling between neighboring centers, so-called hopping conduction [7].

The significance of the blurred band edges is that there is no sharp distinction between the bands, but instead partly localized levels are formed. This leads to a conduction by a process, intermediate between impurity band and hopping conduction, in which the propagation of carriers is characterized by a relatively low mobility [7]. The deeper the levels, the more localized their character, until the deep tail states may be considered proper trapping states. Unlike traps in crystalline materials, however, these deep states would not possess any clearly defined activation energy, since excitation may occur in any one of a wide spectrum of more or less localized levels in the broad zone of the blurred band edge.

A mechanism that tends to favor an aggregated structure is due to the large surface tension forces that exist in solids. The optical behavior observed in certain evaporated films lends further support to the idea of an aggregated rather than a continuous structure [3]. Transparent films of zinc and cadmium deposited at temperature of liquid air are found to possess highly reflecting surfaces. On warming to a room temperature, the transmittance of these films was found to increase. The reflectance decreases and the films show considerably greater scatter than what is seen at the low temperature. This behavior is consistent with the presence of considerable aggregation at the higher temperature, resulting in large interstices between crystallites. The scattering suggests a crystallite size, which is not small compared with the light wavelength used. Further evidence for a discontinuous structure springs from the measurements of the conductivity of thin metal films. Below a limiting thickness, the conductivity is practically zero, suggesting the presence of small crystallites, which are sufficiently isolated to afford no continuous conducting path along the film [3].

3.4 Absorption Spectra of Amorphous Solids

The study of absorption spectra provides a strong tool in understanding the theory of electronic transport in all types of solid-state structures [4]. It is understood that the disorder encountered in amorphous materials gives rise to localized defect states at the edges of valence and conduction bands. In other words, the transitions between these localized states are optically induced. Therefore, the spectra for optical absorption bands can be measured and analyzed for such transitions. It has been noted that the absorption edge curves of many amorphous solids exhibit similar shapes [2]. The absorption edge of such materials can be recognized by three major regions, shown in Figure 3.3.

The behavior in region A arises directly from defect states transitions. The absorption edge extending in regions A and B is rather complicated and contains defect-induced tail at the lowest energies, an exponential region at intermediate energies, and a power-law at the highest energies. The high absorption coefficient $[\alpha(v) > 10^4 \text{ cm}^{-1}]$ in region A is caused by the transitions between the extended states. This is then followed by an exponential region B, which extends over 4 orders of magnitude of $\alpha(v)$.

It has been suggested by Tauc [9] that the exponential absorption edge in interband absorption arises from the electronic transitions between the localized states in the band edge tails. The density of these localized states is assumed to fall off with energy, giving rise to the exponential absorption edge characteristics. This theory has been questioned with regard to the unlikely similarities observed in the slope of the absorption edges of a variety of materials [4]. Instead, Mott and Davis [4] have suggested that the field broadening theory of an exciton line proposed by Dow and Redfield [10] is the most appropriate model to explain the absorption edge. However, both of these theories show close correlation to Urbach's mathematical relation for the absorption coefficient $\alpha(v)$ as a function of photon energy hv [11]:

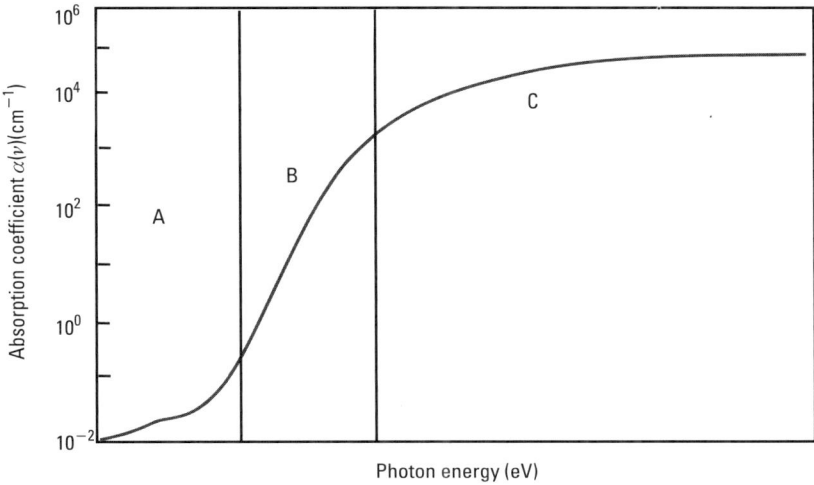

Figure 3.3 An absorption spectrum of amorphous solids consisting of three major regions of interest.

$$\alpha(\nu) = \exp(h\nu/\Delta E) \quad (3.1)$$

where ΔE is the width of the tail of localized states, $\alpha(\nu)$ is the absorption coefficient for the given photon frequency ν, and h is Planck constant. The values of ΔE are calculated from the slopes of the straight lines of $\ln\alpha$ versus $h\nu$ plots.

At high absorption coefficient, Tauc [9] and Mott and Davis [4] proposed the following:

$$\alpha(\nu)h\nu = B(h\nu - E_{opt})^n \quad (3.2)$$

where α is the absorption coefficient, E_{opt} is the optical energy band gap, $h\nu$ is the energy of the incident photons, and B is a constant. The exponent n of the energy dependence of the optical band gap can distinguish four cases of electronic transitions, which are summarized as follows: $n = 1/2$ for direct allowed transition, $n = 3/2$ for direct forbidden transition, $n = 2$ for indirect allowed transition, and $n = 3$ for indirect forbidden transition [4]. Constant B is given by:

$$B = \frac{4\pi\sigma_0}{nc\Delta E} \quad (3.3)$$

where σ_0 is the electrical conductivity at $1/T = 0$, n is the refractive index, c is the speed of light, and ΔE is the width of the tail of the localized states.

A model developed for tails of localized states for ideal chalcogenide glasses is shown in Figure 3.4. Davis and Mott [4] proposed that the tails extend only a few tenths of an electron volt into the forbidden energy band gap, with narrow nonoverlapping at the conduction and valence bands [4]. The two deep levels of localized states, represented by E_X (acceptor) and E_Y (donor), are due to dangling bonds that may be positively or negatively charged.

The optical properties of a thin film generally differ from those of the bulk. The differences are generally attributed to the microstructure of the films [2]. A limited number of momentum vectors is available along the thickness dimension of the film. Therefore, momentum-conserved direct optical transitions are only possible in ultrathin films, in contrast to the direct optical transitions, which can take place in the bulk [2]. Thus, the absorption is reduced in thinner films.

The optical properties of a metal film depend ultimately on the behavior of the conduction electrons in the metal in response to the incident electromagnetic radiation. The adsorption of the electromagnetic waves in a metal is due in part to free electrons and in part to bound electrons. In certain regions of the spectrum, one or another of these sources of absorption may be dominant. The contribution of free electron absorption in films consisting of aggregates will be considerably higher than that in the bulk metal, due the much larger extent of scattering at the particle boundaries [3]. The contribution of the bound electrons would be expected to change with the state of aggregation.

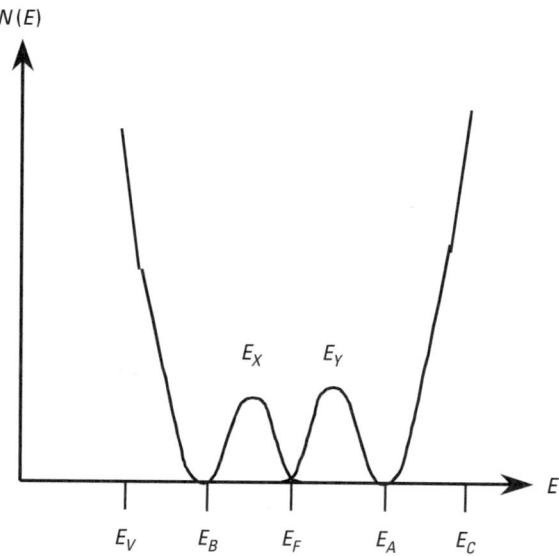

Figure 3.4 Davis and Mott model.

3.5 Metal-Semiconductor Contacts

In order to measure the electrical conductivity of a thin or thick film, it is necessary to bring the metal or the semiconductor electrodes into contact with the surface of the material of interest. This allows either injection or withdrawal of electrons from the bulk of the material. It was proposed that the charge carriers generated due to insulator-metal interface are the measure of the electrical conductivity of the dielectric [12]. The role of the semiconductor is to create a potential barrier between the electrodes. Such a potential barrier determines precisely the predominance of a certain conduction mechanism over others at the metal-insulator interface.

Metal-semiconductor contacts are an obvious component of any semi-conductor device. At the same time, such contacts cannot be assumed to be as low resistance as that of two connected metals. In particular, a large mismatch between the Fermi energy of the metal and semiconductor can result in a high-resistance rectifying contact. A proper choice of materials can provide a low-resistance ohmic contact [13]. There are three types of metal-semiconductor contacts: ohmic, blocking, and neutral contacts. The type of contact formed in any given system may be determined by considering the relative values of the work functions of the metal φ_m and the semiconductor φ_s, and also by the thermal equilibrium conditions.

3.5.1 Ohmic Contact

If φ_s is greater than φ_m and if thermal equilibrium is satisfied, then electrons are injected from the electrode into the conduction band of the semiconductor. This results in a space charge storage, which acts as a reservoir of charge, capable of supplying electrons for conduction as required by the bias demands [13]. For an n-type semiconductor, the work function of the metal φ_m must be close to or smaller than the electron affinity of the semiconductor. For a p-type semiconductor, the work function of the metal φ_m must be close to or larger than the sum of the electron affinity and the band gap energy [13]. This causes the bottom of the conduction band to curve upward and away from the metal-semiconductor interface. The conduction therefore depends on the rate of current flow through the semiconductor material. The process, therefore, is limited by the resistivity of the semiconductor, and the conduction mechanism is known as bulk limited. Figure 3.5 illustrates semiconductor band diagram for an ohmic contact.

3.5.2 Neutral Contact

If φ_m is equal to φ_s, as shown in Figure 3.6, there is no accumulation of charge on either side of the metal-semiconductor interface. As a result, there is neither

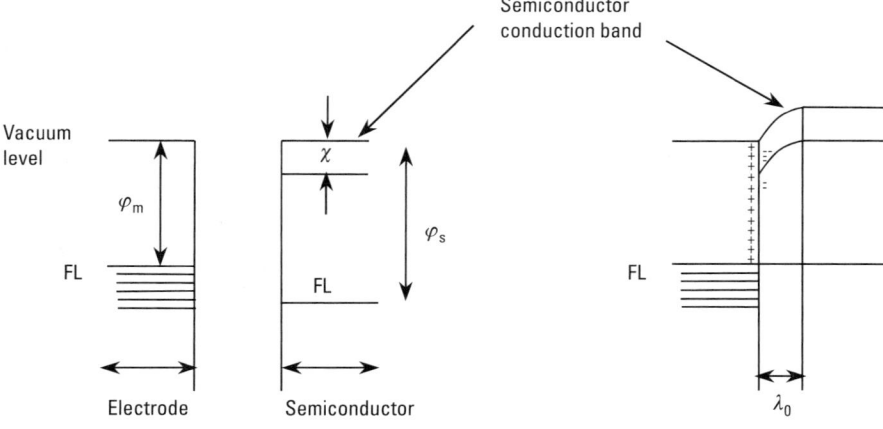

Figure 3.5 Ohmic contact, for which $\varphi_m < \varphi_s$.

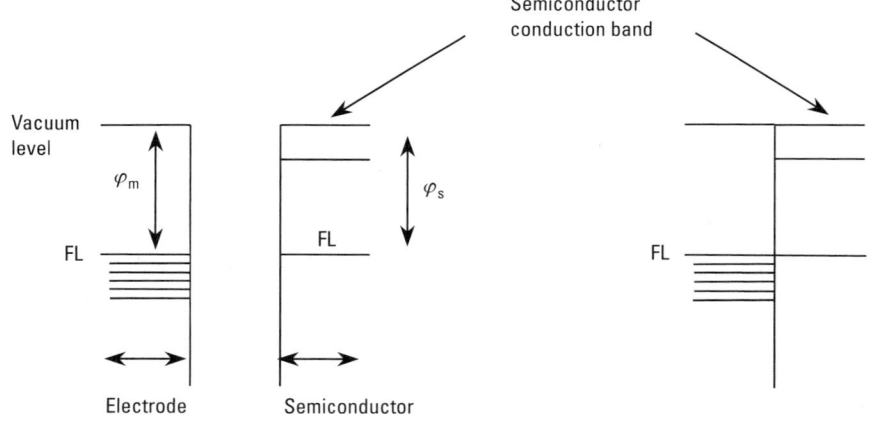

Figure 3.6 Neutral contact, for which $\varphi_m = \varphi_s$.

charge transfer nor band bending [13]. This type of contact represents the transitional stage between ohmic contacts and blocking contacts.

3.5.3 Blocking Contact

When φ_m is greater than φ_s, a blocking contact occurs, and the electrons flow from the semiconductor into the electrodes in order to establish thermal equilibrium conditions (i.e., for the Fermi levels to align). Because of this, a positive space charge region is created in the semiconductor and a negative space charge region resides in the metal electrode [13]. The electrostatic interaction between the positively and negatively charged regions leads to the creation of a local field within the surface of the semiconductor. As a result, the bottom of the conduction band of the semiconductor curves downward in the metal-semiconductor interfaces. The rate of electron flow is limited by crossing over the potential barrier. This process is known as electrode limited and is illustrated in Figure 3.7.

3.6 Conduction Mechanisms in Amorphous Materials

The mechanism of ionic conduction involves the drift of ions or vacancies under the influence of an electric field by hopping over a potential barrier ϕ from one direct site to the next neighboring site. The high density of the structural defects in amorphous dielectric films leads to a high activation energy, a large transit time, transportation of materials from one electrode to another, and the polarization effects in a direct current field to be the dominant features of the ionic conduction [4].

Impurity conduction is a process in which the electron moves between the centers without activation into the conduction band. The process involves an electron, occupying an isolated donor level, and has a wave function localized slightly below the conduction band minimum. This type of conduction may be observed in an insulator since it has a very low density of thermally generated carriers in the conduction band, rather than in semiconductors such as Si and Ge. In any amorphous material, for the impurity conduction to be significant, it must contain simultaneously donor and acceptor (trap) centers. One may regard the donors as the source of carriers and the traps as the medium through which

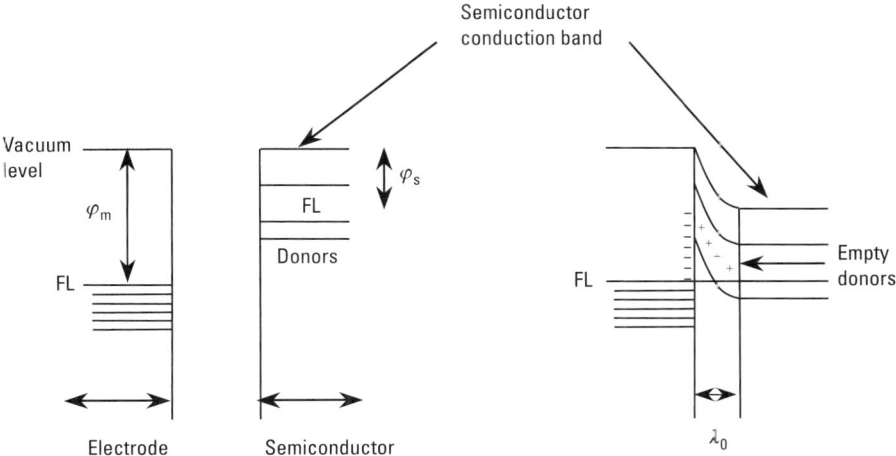

Figure 3.7 Blocking contact, for which $\varphi_m > \varphi_s$.

they move. Without the presence of acceptors, impurity conduction is not possible until the interaction between centers is very large (high concentration of impurities), as when a type of metallic conductivity sets in. For lower concentrations, the electrons move by a hopping process from one center to another.

Electronic conduction of electric current through dielectric films, especially if they are of amorphous or glassy structure, may be due to the motion of the free carriers—electrons in the conduction band or holes in the valence band—or to the motion of quasi-localized carriers. This is described as hopping of bound carriers between the localized states in the dielectric [7]. The former process requires activation energy in order to excite a carrier into the relevant band. This energy can normally be supplied thermally or by other free carriers, which have acquired a high energy in an electric field, leading to avalanche process, as in the Poole-Frenkel effect. The hopping process requires less energy than the activation into the free band, and this energy may, in the limit of the very high density of localized centers, tend to zero, as in the case of impurity band conduction in semiconductors [7]. This process is favored in the case of heavily disordered solids, such as amorphous and glassy dielectric films. Some dielectric films show a region of linear voltage-current characteristic (i.e., ohmic conduction at low fields), especially at elevated temperatures. It is difficult to establish conclusively whether ohmic conduction, when present, is due to ionic or electronic processes, but ionic conduction would appear more likely [7].

It is natural to expect the presence of hopping conduction in highly disordered and amorphous materials. The salient features of hopping conduction may be summed up as follows [7]:

1. Linear dependence of current on voltage;
2. Relatively low activation energy, much lower than the activation energy of the relevant donor or acceptor center;
3. Monolithic increase of alternating current conductivity with frequency;
4. Very slight decrease of polarizability with frequency.

In noncrystalline materials, it is not possible to make a detailed estimate of the magnitude of the activation energy for hopping. However, in materials with hopping conduction mechanism, one could expect saturation or even decrease of conductivity at high frequencies. This limiting process would be expected at frequencies comparable with the natural transition frequency for hopping. The implication is that the natural hopping frequency must lie at least in the microwave region, if not in the infrared region of the spectrum. When referring to power-law dependence of current upon voltage in the context of dielectric films, one naturally tends to think of space-charge-limited flow in solids. However, power-law relations with high exponents that extend over many decades of current are frequently observed in amorphous and glassy materials in a wide range of temperatures [7].

The distinction between electronic and ionic conductions may be a very difficult problem, and sometimes there is no conclusive answer. One important criterion of ionic conduction is material transport, which should be detectable if sufficiently high currents are passed for long enough. As well, slow polarization may result and be detectable as current falling with time and as standing potential on open circuit [7]. In crystalline materials, ionic conduction is characterized by low mobilities and high activation energies, while electronic conduction is associated with relatively higher mobilities and low activation energies. In the case of amorphous materials, mobility may be very low, so for hopping conduction, the activation energy should be small. It is not possible to give definite limits for activation energy, but as a general rule values less that 0.2 eV should be considered as indicating undoubtedly electronic mechanisms, and values in excess of 0.6–0.8 eV would normally be associated with ionic transport [7]. Hopping processes normally fall in category of electronic processes, and substantially higher activation energies are possible.

The transitions processes determining the dielectric response of the semiconductors could be of the following types [14]:

1. Excitation from shallow donors into the conduction band—with typical energies in the meV range and the corresponding acceptor-valence band excitations. These are extremely rapid at any temperatures, except cryogenic, and they produce free carriers, which are freely mobile in their respective bands. Thus, their response may be considered to be instantaneous, and therefore leading to purely direct current transport up to frequencies comparable to the collision frequency, which is of the order of 10^{14} s^{-1}.

2. Vertical transition involving relatively large changes of energy. In this transition, the excitation takes place from deep traps into the bands with corresponding reverse capture processes. The energy dissipation in downward transitions may be through phonons or photons. The energy needed for upward transitions has to come from phonons in absence of external illumination, and this may entail considerable time delays.

3. Horizontal transitions involving relatively small changes of energy, such as hopping near Fermi level, especially in relatively shallow levels at low temperatures.

4. Strictly localized hopping of an electron or an ion in a potential double well, arising from tight binding at some localized defect. Although the energies involved are quite small, the frequency dependence is

found to extend down to quite low frequencies, which suggests that considerable delays are involved [14].

3.6.1 Schottky Versus Poole-Frenkel Conduction Mechanism

If the potential barrier is too thick to allow tunneling to take place or at sufficiently high temperature, electrons can be thermally excited to cross over a barrier (ϕ) into the conduction band of the insulator. The thermal emission current density is given by the Richardson-Schottky relation:

$$J_s = AT^2 \exp(-\phi/kT) \tag{3.4}$$

where $A = 4\pi emk^2/h^3$ and J_s is the Shottky current density. When an electric field is applied, the potential barrier is lowered by $\Delta\varphi$:

$$J_s = AT^2 \exp\left(\frac{-\phi - \beta_s E^{1/2}}{kT}\right) \tag{3.5}$$

where E is the applied electric field and β_s – is the Schottky field-lowering coefficient, expressed by:

$$\beta_s = \left(e^3/4\pi\varepsilon_r\varepsilon_0\right)^{1/2} \tag{3.6}$$

where ε_0 and ε_r are the free space and relative permittivities, respectively.

In Poole-Frenkel process, the thermally excited electrons are driven by the applied electric field to move from localized traps to the conduction band. This process is also known as field-assisted thermal ionization, since the emission of electrons occurs from trapping centers in insulators by the joint effect of temperature and electric field. Poole-Frenkel field-lowering coefficient (β_{pf}) can be written as:

$$\beta_{pf} = \Delta\varphi/E^{1/2} = \left(e^3/\pi\varepsilon_0\varepsilon_r\right)^{1/2} \tag{3.7}$$

The Poole-Frenkel coefficient in a uniform electric field is twice than that due to Schottky effect:

$$\beta_{pf} = 2\beta_s \tag{3.8}$$

By considering a number of factors, it is possible to discriminate between the two types of emission processes. For example, the contact barrier height for insulators having large energy band gaps is expected to be greater than 0.8 eV, whereas for bulk emission, lower activation energy may be found.

3.7 Radiation Damage in Crystalline Structures

The interaction of different incident particles with the electrons of the crystalline lattice leads to a change of electron energy. The change is not permanent, and the thermal equilibrium of the electron within the crystal lattice is restored in a few milliseconds. This is not the case if the particle energy is transferred to an atomic nucleus. If the imparted energy is less than a binding energy of the nucleus in the lattice, the atom is elastically displaced from its site and the displacement is not permanent. If the transferred energy is higher than the binding energy of the atom, it is permanently removed from the site. This results in a vacancy at the original site with simultaneous appearance of a surplus atom in an interstitial position between the atoms. This is a Frenkel-type lattice effect. The vacancy behaves like an acceptor; the surplus atom behaves like a donor. Therefore, new energy levels appear in the crystal under the action of radiation.

The change in conductivity caused by the new levels is difficult to estimate, since neither the energy nor the number of new levels is accurately known. The ionizing energy of new levels is high, and the low-lying levels may carry out the donor as well as the acceptor impurity effects due to this compensating effect. In the initial stage of irradiation, the resistance increases in any material containing donors or acceptors of low ionizing energy. If the irradiation is continued after the compensation of the impurities present in the original crystal, the formation of acceptor levels is permanent. The lattice defects affect more than the conductivity of the crystals. The deep-lying energy levels produced by irradiation are very effective recombination centers, which can reduce the carrier lifetimes by several orders of magnitude. These lattice defects generated by radiation damage last for several months at room temperature.

Bombarding particles of small mass transfer only a small fraction of their energy to the atomic nucleus of semiconductors, and the energy transfer takes place in several steps. The probability of nuclear interaction, particularly for particles having energies above a few hundred keV, is much lower than that of interaction leading to ionization of electron shell and excitation. The latter interactions do not cause lattice defects to the crystal. Calculations show that 1 MeV of protons and α-particles impart 200 Frenkel defects in Si. In the case of electrons colliding with an atomic nucleus, electrons with energies below 1 MeV do not produce, as a rule, any lattice defects, while those with energies from 1 to 10 MeV cause from 1 to 10 lattice defects in Si. If the crystal is exposed to γ-radiation, electrons produced by Compton interaction may damage the lattice.

Fast neutrons collide classically with the atomic nucleus. These neutrons do not lose energy to ionization. Thermal neutrons cannot directly transfer as much energy to the crystal as is needed for the removal of an atom. The capture of a thermal neutron by an atom results in a nuclear reaction, by which charged particles and gamma quanta are produced. The capture of a thermal neutron by the ^{28}Si and ^{29}Si isotopes leads a pure reaction. If a thermal neutron is captured by ^{30}Si of about 4% abundance in silicon, it changes to β-active ^{31}Si, which decays to ^{31}P.

The energy lost by ionizing radiation in semiconductor detectors ultimately results in the creation of electron-hole pairs. The average energy ε, necessary to create an electron-hole pair in a given semiconductor at a given temperature, is independent of the type and the energy of the ionizing radiation. The values of ε are 3.62 eV in silicon at room temperature, 3.72 eV in silicon at 80K, and 2.95 eV in germanium at 80K.

Since the forbidden band gap value is 1.115 eV for silicon at room temperature and 0.73 eV for germanium at 80K, it is clear that not all the energy of the ionizing radiation is spent in breaking covalent bonds. Some of it is ultimately released to the lattice in the form of phonons.

The constant value of ε for different types of radiation and for different energies contributes to the versatility and flexibility of semiconductor detectors for use in nuclear spectroscopy. The low value of ε compared with the average energy necessary to create an electron-ion pair in a gas (typically 15 to 30 eV) results in superior spectroscopic performance of semiconductor detectors. In general, radiation damage reduces the lifetime of minority carriers [15, 16].

3.8 Radiation-Induced Defects in Oxide Materials

Properties of metal oxides materials are directly or indirectly connected to the presence of defects, oxygen vacancies in particular [17, 18]. These defects determine the optical, electronic, and transport properties of the material and usually dominate the chemistry of its surface. Oxygen vacancies are naturally present in every oxide in the form of Shottky or Frenkel defects, and their concentration can be increased or reduced in several ways [18]. Point defects play a fundamental role in determining the physical and chemical properties of inorganic materials. This holds not only for bulk properties, but also for the surface of oxides, where several kinds of point defects exist and exhibit a rich and complex chemistry. Depending on the electronic structure of the material, the nature of oxygen vacancies changes dramatically. Examples include nonmetal vacancy at nonmetal site, metal vacancy at metal site, neutral vacancies, positively/negatively charged nonmetal vacancies, and free positive holes [19].

Oxygen vacancies are known as color centers, or F centers (from *Farbe*, the German word for color). If the vacancy is localized at the surface, a subscript *s* is added, F_s. F centers exist in three states, depending on their electronic charge. Diamagnetic oxygen vacancies can be either neutral (F) or doubly charged (F^{2+}). In the former case, two electrons are associated with the vacancy. Of particular interest for the characterization of oxygen vacancies are the paramagnetic F^+ centers, consisting of a single electron trapped in the cavity formed by removing an oxygen anion, O^- [18].

It is believed that ionizing radiation causes structural defects, leading to a change in density upon exposure to γ-rays [20]. The energy required to create an electron-hole pair is relatively small (e.g., 18 eV for SiO_2) [1]. The model for color center kinetics assumes that the level of the radiation damage should be dose-rate dependent because of the damage recovery [21]. As color centers are created under irradiation, they also annihilate even under room temperature. During irradiation, both annihilation and creation coexist. The color center density will reach equilibrium at a level depending on the applied dose rate. The creation and annihilation constants can be determined by using experimental data obtained under one particular dose rate and can then be used to predict the behavior of the same sample under different dose rates [21].

Color centers in oxide thin films, such as WO_3 and MoO_3, have been observed by irradiation with UV light from a high-pressure lamp in the fundamental absorption region at a wavelength of 330 nm [22]. The formation of color centers has been associated with an increase in electrical conductivity, in which free electrons are produced as a result of band-to-band transitions and trapping of these electrons in oxygen ion vacancies. It was found that there are more than one type of defect responsible for the formation of color centers in MoO_3 thin films [22]. Alternatively, both sputtered SiO_2 thin films and fused SiO_2 are identical in producing color centers when they are bombarded by neutrons or X-rays [23]. It is generally accepted that two distinct processes are responsible for the formation of color centers, following bombardment with ionizing radiation. The primary mechanism is in charge for defect formation, while the secondary one gives rise to the stabilization of the centers [24]. The overall response is roughly considered three-staged, where initial formation of color centers (stage I) is followed by defects saturation (stage II) and concluded with a decrease in defect concentration (stage III). The primary process, which is active during all three stages, involves the nonradiative de-excitation of an exciton with the formation of a Frenkel pair [25] (F center and halogen interstitial). Preexisting impurities and defects in the sensing material act as saturating traps, capable of stabilizing a limited number of interstitial atoms. According to that, the concentration growth occurring during stage I is due to the effective stabilization of the interstitial ions by the traps initially present in the crystal. The saturation of the traps leaves the recombination of the interstitial atoms and

F centers the dominant process and leads to the flat region of stage II. During both stages, the contribution to the stabilization process due to the aggregation of the interstitial atoms is still negligible, although it becomes more significant as the radiation dose increases. The interstitial aggregation finally dominates the stabilization mechanism when, during stage III, it allows a further increase in the defect concentration.

3.9 Radiation Effects in Polymers

Radiation interacts with polymers in two ways: chain scission, which results in reduced tensile strength and elongation, and crosslinking, which increases tensile strength but reduces elongation. Both reactions occur simultaneously, but one is usually predominant, depending on the specific polymer and additives involved. Chain scission classically affects stressed polymers (containing residual molding stress) to a greater extent than nonstressed polymers. The combined impact of solvent-induced stress, residual molding stress, and applied load acts to intensify radiation damage. Generally, polymers containing aromatic ring structures (e.g., polystyrene) are resistant to radiation effects. Aliphatic polymers exhibit degrees of resistance, depending on their levels of unsaturation and substitution. A linear correlation of polymer conductivity with the applied dose was reported [26, 27]. Two nonconducting forms of polyaniline were used as a detecting device for low-dose ionization radiation, whereas the conducting form manifested itself as a possible high-dose detector [27]. The radiation interaction was mainly due to an oxidation process of the main polymer chain yielding a conductivity alteration. Ionizing radiation also induced a doping state in the polymer main chain structure similar to that found in the conventional acid-doping process [27].

Table 3.1 provides an overview of the polymers commonly used for medical devices, along with their typical characteristics following irradiation. It is important to remember that not all brand products share the same characteristics. For some materials and products that are sensitive to oxidative effects, such as low molecular weight polypropylene, polytetrafluorethylene, and polyacetals, radiation tolerance levels for electron beam exposure may be slightly higher than for gamma exposure. This is due to the higher dose rates and shorter exposure times of e-beam irradiation, which reduces the degradative effects of oxygen. A comparison of radiation's effect on e-beam versus gamma is not easily accomplished unless product-specific characteristics—including part thickness, volume of product, molecular weight, scission to crosslink ratio, oxygen sensitivity, use of antioxidants, and aging effects—are known and entered into the evaluation.

Some effects of radiation, such as reduced elongation due to chain scission, may detract from the device's performance. Others can be beneficial. For example, crosslinking of polyethylene and silicones increases tensile

Table 3.1
Radiation Tolerance Levels of Polymers Used for Medical Applications

Material	Tolerance Level (kGy)	Comments
Butyl	50	Sheds particulate after irradiation
Ethylene-propylene diene monomer (EPDM)	100–200	Crosslinks, yellows slightly
Fluoro elastomer	50	Avoid multiple sterilization
Natural rubber (isoprene)	100	Very stable with sulfur or resin cure systems; avoid stressing product by not bending, folding, or wrinkling in packaging
Nitrile	200	Avoid multiple sterilization
Polyacrylic	50–200	Avoid multiple sterilization
Polychloroprene (neoprene)	200	Avoid multiple sterilization
Silicones (peroxide and platinum catalyst system)	50–100	Crosslink density increases more in peroxide systems than in platinum systems; silicones retain a slight memory of coiling shape in packaging
Styrene-butadiene	100	Avoid multiple sterilization
Urethanes	100–200	Wide variations in urethane chemistry applied to medical device; requires testing of part process and geometry to validate
Allyl digylcol carbonate (polyester)	5,000–10,000	Retains clarity
Epoxies	1,000	Many good formulations available; test the formulation selected for use; frequently substituted for toxic solvents in assembly; success depends on joint design and application process
Phenolics	50,000	
Polyesters	10–1,000	Use of glass and other fillers improves physicals
Polyurethanes	100–1,000	Wide formulation variations for urethanes; dose tolerance depends on monomers used in formulation
Acrylonitrile/butadiene/styrene (ABS)	1,000	Protected by benzene ring structure; butadiene impact modifier degrades above 100 kGy; avoid high dose on high-impact grades

Table 3.1 (continued)

Aromatic polyesters (PET, PETG)	1,000	Very stable, retains excellent clarity; drying is essential; good in luer connectors
Esters and ethers	50	Thin films and fibers embrittle above 50 kGy
Paper, card, corrugated fibers	100–200	Papers discolor and become brittle, but are acceptable for single use
Cellulose, acetate, propionate, and butyrate	50	Plasticized versions slowly become brittle above 50 kGy
Tetrafluoroethylene (PTFE)	5	Liberates fluorine gas, disintegrates to powder; avoid use
Polychlorotrifluoroethylene (PCTFE)	200	
Polyvinylidene fluoride (PVDF)	1,000	
Ethylene-tetrafluoroethylene (ETFE)	1,000	
Fluorinated ethylene propylene (FEP)	50	
High-performance engineering resins	1,000–10,000	Substitutes for metal, these resins have high strength and good elongation that tolerate radiation well; resins include nylon, polycarbonate, ABS, polysulfone, polyester, polyether ketone, liquid crystal polymer, polyetherimide, and polyimide
Polyacetals (delrin, celcon)	15	Avoid use due to embrittlement
Polymethylmethacrylate	100	Yellows at 20–40 kGy; clarity recovers partially on aging
Polyacrylonitrile	100	Yellows at 20–40 kGy
Polyacrylate	100	Yellows at 20–40 kGy
Polycyanoacrylate	200	Many good formulations; adhesives function at 100 kGy with less than 30% degradation
Aliphatic and amorphous grades	50	Discolors, no resterilization; avoid thin films and fibers; nylon 11 and 12 perform better; dry before molding
Aromatic polyamide-imide	10,000	High heat/strength grade; stabilized by benzene ring structure.
Polycarbonate	1,000	Discolors, clarity recovers after aging; dry before molding

Table 3.1 (continued)

Polyethylene (LDPE, LLDPE, HDPE, UHMWPE)	1,000	Crosslinks to gain strength, loses some elongation; all polyethylenes tolerate radiation well; low density is most resistant; HDPE packaging film including spin-bonded porous packaging may lose 40% to 50% elongation at doses of 50 kGy; implants of UHMWPE have reports of early wear at 50 kGy
Polyamides	10,000	
Polymethylpentene	20	Subject to oxidation degradation; avoid use
Polyphenylene sulfide	1,000	
Polypropylene, radiation stabilized		Higher tolerance levels reported using e-beam
Homopolymer	20–50	Used with marginal success in syringes; subject to orientation and oxidation embrittlement; degrades over time; validate with real-time aging; avoid use of nonstabilized polypropylene
Copolymers of propylene-ethylene	25–60	More stable than homopolymer; successful in syringe applications using suitable stabilizer package
Polystyrene	10,000	All styrenes are stabilized by benzene ring structure
Polysulfone	10,000	Amber color before irradiation
Polyurethane, polyether, and polyester (rigid and flexible)	100–200	Excellent physicals and chemical resistance to stress cracking; drying is essential to success; good in luer connectors; all types show irreversible yellowing
Polyvinylbutyral	100	Yellows
Polyvinylchloride (PVC)	100	Yellows, can be tinted for color correction; success depends on quality of material, formulation, and processing; tubing crosslinks becoming slightly stiffened
Polyvinylidene chloride (PVDC)	100	Yellows, releases HCl
Styrene/acrylonitrile (SAN)	1,000	Yellows at 40 kGy

Source: [28].

strength. Manufacturers should be cognizant of the possible impact of radiation on mechanical properties such as tensile strength, elastic modulus, impact strength, and elongation. Outcomes may influence performance and should be evaluated in advance by functional testing.

3.10 Radiation-Induced Degradation Processes in Device Parameters

The effect of irradiating an electronic material and the consequent degradation in performance of devices made from such material can follow a number of events. The final result will depend on the type of radiation, its mode and rate of interaction with the materials, the type of materials, its particular contribution to the device function, and the physical principles upon which the function of the device is based [1]. Figure 3.8 summarizes the radiation-induced degradation effects in solid-state devices and materials.

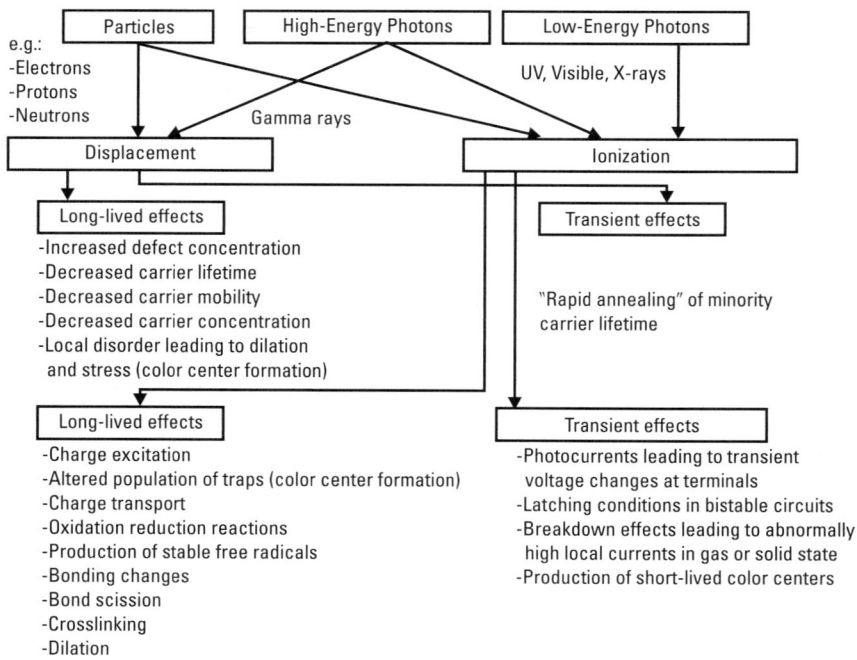

Figure 3.8 Summary of radiation-induced degradation effects. (*From:* [1]. © 1993 Oxford University Press. Reprinted with permission.)

For the practical design of suitable sensors, sample geometries and operation modes have to be found, which favor only one of the different contributions to the overall conductivity in a certain sensor material [19]. This contribution should contain in the best case predominantly the changes of only one type of external exposure.

In the light of this information concerning the types of radiation dosimeters in current use, one may see that there is a need for a reliable, cost-effective sensor for personal dosimetry application. Sensors fabricated using thin and thick film technologies could be regarded as a basis for the desired end product. These are discussed in detail in Chapter 4.

References

[1] Holmes-Siedle, A. G., and L. Adams, *Handbook of Radiation Effects,* New York: Oxford University Press, 1993.

[2] Chopra, K. L., *Thin Film Phenomena,* Malabar, FL: Robert E. Krieger, 1979.

[3] Heavens, O. S., *Optical Properties of Thin Solid Films,* New York: Dover, 1991.

[4] Mott, N. F., and E. A. Davis, *Electronic Process in Non-Crystalline Materials,* Oxford, U.K.: Clarendon, 1979.

[5] Mott, N. F., "Conduction in Non-Crystalline System," *Phil. Mag,* Vol. 19, 1969, pp. 8–35.

[6] Mott, N. F., *Conduction in Non-Crystalline Materials,* Oxford, U.K.: Clarendon, 1997.

[7] Jonscher, A. K., "Electronic Properties of Amorphous Dielectric Films," *Thin Solid Films,* Vol. 1, No. 3, 1967, pp. 213–234.

[8] Arshak, K. I., and C. A. Hogarth, "Effects of Annealing on the Structure, Electron Spin Resonance and Optical Energy Gap of Thin BaO-SiO Films," *Thin Solid Films,* Vol. 137, 1986, pp. 281–291.

[9] Tauc, J., *Optical Properties of Solids,* New York: Academic, 1966.

[10] Dow, J. D., and D. Redfield, "Electroabsorption in Semiconductors: The Excitonic Absorption Edge," *Phys. Rev. B,* Vol. 1, No. 8, 1970, pp. 3358–3371.

[11] Urbach, F., "The Long Wavelength Edge of Photographic Sensitivity and of the Electronics of Solids," *Physical Review,* Vol. 92, 1953, p. 1324.

[12] Maissel, L. I., and R. Glang, *Handbook of Thin Film Technology,* New York: McGraw-Hill, 1983.

[13] Hunter, L. P., *Handbook of Semiconductor Electronics,* New York: McGraw-Hill, 1970.

[14] Jonscher, A. K., "Dielectric Characterisation of Semiconductors," *Solid-State Electronics,* Vol. 33, No. 6, 1990, pp. 737–742.

[15] Iles, P. A., "Evolution of Space Solar Cells," *Solar Energy Materials and Solar Cells,* Vol. 68, No. 1, 2001, pp. 1–13.

[16] Kishimoto, N., et al., "Radiation-Induced Conductivity of Doped Silicon in Response to Photon, Proton and Neutron Irradiation," *Journal of Nuclear Materials,* Vol. 283–287, No. 2, 2000, pp. 907–911.

[17] Tilley, R. J. D., *Principles and Applications of Chemical Defects,* Cheltenham, U.K.: Stanley Thornes, 1998.

[18] Pacchioni, G., "Ab Initio Theory of Point Defects in Oxide Materials: Structure, Properties, Chemical Reactivity," *Solid State Sciences,* Vol. 2, No. 2, 2000, pp. 161–179.

[19] Gopel, W., and G. Reinhardt, "Metal Oxide Sensors: New Devices Through Tailoring Interfaces on the Atomic Scale," in *Sensors Update,* H. Baltes, W. Gopel, and J. Hesse, (eds.), New York: VCH, 1996, pp. 49–120.

[20] Zhu, R. Y., "Radiation Damage in Scintillating Crystals," *Nuclear Instruments and Methods in Physics Research Section A: Accelerators, Spectrometers, Detectors and Associated Equipment,* Vol. 413, No. 2–3, 1998, pp. 297–311.

[21] Deng, Q., Z. Yin, and R. Y. Zhu, "Radiation-Induced Color Centers in La-Doped PbWO4 Crystals," *Nuclear Instruments and Methods in Physics Research Section A: Accelerators, Spectrometers, Detectors and Associated Equipment,* Vol. 438, No. 2–3, 1999, pp. 415–420.

[22] Chopoorian, J. A., G. H. Dorion, and F. S. Model, "Photochromism of Metal Oxides—The Light Sensitivity of MoO_3 or WO_3 Coprecipitated with TiO_2," *Journal of Inorganic and Nuclear Chemistry,* Vol. 28, No. 1, 1966, pp. 83–88.

[23] Pliskin, W. A., R. G. Simmons, and R. P. Esch, *Thin Film Dielectrics,* New York: Electrochemical Society, 1969.

[24] Baldacchini, G., et al., "Selective Production of Aggregate Centers in LiF Crystals by Ionizing Radiations," *Nuclear Instruments and Methods in Physics Research Section B: Beam Interactions with Materials and Atoms,* Vol. 141, No. 1–4, 1998, pp. 542–546.

[25] Crawford, F. H., *Thermodynamics for Engineers,* New York: Harcourt: Brace & World, 1968.

[26] Arshak, K., et al., "Thin and Thick Films of Metal Oxides and Metal Phthalocyanines as Gamma Radiation Dosimeters," *IEEE Trans. on Nuclear Science,* Vol. 51, No. 5, 2004, pp. 2250–2255.

[27] Lima Pacheco, A. P., E. S. Araujo, and W. M. de Azevedo, "Polyaniline/Poly Acid Acrylic Thin Film Composites: A New Gamma Radiation Detector," *Materials Characterization,* Vol. 50, No. 2–3, 2003, pp. 245–248.

[28] Sterigenics International, http://www.sterigenics.com.

4

Gamma Radiation Dosimetry Using Metal Oxides and Metal Phthalocyanines[1]

4.1 Thin and Thick Film Technologies

4.1.1 Thin Film Technologies

There are a wide variety of techniques for deposition of thin films. Examples are thermal evaporation (also known as vacuum vapor deposition), electron-beam evaporation, magnetron sputtering, chemical vapor deposition (CVD), and molecular beam epitaxy (MBE) [1–4]. Vacuum evaporation is a method used to deposit many types of materials in a highly evacuated chamber, in which a material is heated by electricity. It consists of vaporizing a solid material by heating it to sufficiently high temperatures and recondensing it onto a cooler substrate to form a thin film. A large current passes through a filament container (usually in the shape of a basket, boat, or crucible) with finite electrical resistance, thus heating the material. The evaporation temperature and its inertness to alloying/chemical reaction with the evaporant dictate the choice of this filament material. This technique is also known as indirect thermal evaporation, since a supporting material is used to hold the evaporant. Once the metal is evaporated, its vapor undergoes collisions with the surrounding gas molecules inside the evaporation chamber. As a result, a fraction is scattered within a given distance during their transfer through the ambient gas. The mean free path for air at 298K is approximately 45 cm and 4,500 cm at pressures of 10^{-4} and 10^{-6} torr, respectively. Therefore, pressures lower than 10^{-5} torr are necessary to ensure a straight-line

---

1. In cooperation with A. Arshak and S. Zleetni.

path for most of the evaporated species and for substrate-to-source distance of approximately 10 cm to 50 cm in a vacuum chamber. The substrates with appropriate masks are placed above and at some distance from the material being evaporated. When the process is completed, the vacuum is released and the masks are removed from the substrates. This process leaves a thin, uniform film of the deposition material on all parts of the substrates exposed by the open portions of the mask.

The vacuum evaporation technique is most suitable for deposition of highly reactive materials, such as aluminum, that are difficult to evaporate in air. The method is clean and allows a better contact between the layer of deposited material and the surface upon which it has been deposited. In addition, because evaporation beams travel in straight lines, very precise patterns may be produced. In general, thermal vacuum deposition produces films with structural defects, such as grain boundaries or lattice imperfections [4]. The so-called minor defects, which are frequently observed in deposited films, include dislocation loops, stacking-fault tetrahedral, and small triangular defects; all of these are generally attributed to vacancy collapse [4]. Controlling the deposition conditions such as pressure, deposition rate, substrate temperature, and surface nature can alter the intensity of such defects. The settings for the evaporation procedure vary, depending on the type of the material being deposited and desired film properties, such as thickness and conductivity.

4.1.2 Thick Film Technology

For decades screen-printing has been the dominant method for thick film manufacture due to its low cost. Many models of the printing process have been developed in both research laboratories and industry since the 1960s. With a growing need for denser packaging and a drive for higher pin count, screenprinting has been refined to yield high-resolution prints [5]. Thick film technology offers the designer a wide variety of available materials, flexibility in device design, and easy integration with electronic circuitry and device-packaging materials [6]. Thick film technology involves screenprinting of specially formulated pastes in definite patterns and sequences to produce individual components and complete functional circuits [7]. The subsequent firing cycles, associated with thick film fabrication, mature the printed paste elements and bond them integrally to the substrate. Thick film technology is expected to play an increasing role in the packaging of semiconductor devices and power electronic applications.

One of the key advantages of the thick film process over thin film is its ability to form multilayer interconnect circuits. The process of thick film manufacture permits multiple layers of conductors, dielectrics, and resistors to be screenprinted and fired/cured onto the substrate, resulting in high-density interconnections. The process for firing/curing each additional layer follows the same firing process, whereby the additional layers are printed, dried, and fired/cured on top of the previous layer. This is only possible as long as the following layers do not have peak firing/curing temperatures higher than that of the previous layers, as it would cause uncontrollable diffusion of the materials. Another distinct advantage of thick film processing lies in the ability to cofire/cure certain printed layers in order to significantly shorten the processing time involved, thus making the device more cost effective. However, this is only advised if the cofired/cured layers are passive, such as in dielectrics or layers of the same material. However, passivation of a circuit requires the firing of an extra layer, to which end a low-melting dielectric ink is used.

It is important to remember that each technology has its own limitations with respect to the material that can be used. For example, materials that have very high melting point (more than 1,800°C) cannot be deposited by thermal vacuum evaporation techniques. For thick films, suitable materials should be easily printed and baked.

4.2 Thin Films as Radiation Sensors

The ^{60}Co (1.17 and 1.33 MeV) and ^{137}Cs (0.662 MeV) sources were used to expose the samples to γ-radiation. Values of radiation damage were estimated from changes in the electrical, optical, and structural characteristics of the materials. Methods of analysis included:

- Characterization of the electrical properties of thin/thick films individually and as p-n junctions, capacitive, planar, and so forth, by impedance spectroscopy (Hewlett Packard impedance analyzer HP 4277A LCZ-meter), I-V characteristics, Hall effect, and other relevant techniques;

- Measurement of optical properties using CARY 1E UV-Visible Spectrophotometer;

- Qualitative X-ray powder diffraction (XRD) using a Philips X'pert PRO MPD (Multi Purpose Diffractometer) X-ray diffractometer PW3050/60 $\theta-\theta$ (Philips, Eindhoven, the Netherlands);

- Scanning electron microscope (SEM) analysis of the samples using Jeol JSM-840 SEM;

- Raman Spectroscopy on a DILOR XY Labram using a 20-mW red laser through an 1800 grating where the spectra were collected with a Peltier cooled CCD detector and an excitation source of 632.81 nm.

4.2.1 Metal Oxides Thin Films

4.2.1.1 Defects in Metal Oxides

Metal oxides are widely used in various technological applications, such as coating, catalysis, electrochemistry, optical fibers, and sensors [8, 9]. However, the most important properties of these materials relate to the presence of defects [10]. Such defects can be vacancies, substitutional dopants or impurities, substututional interstitials, or impurities that migrate. These are important because they behave differently in amorphous and crystalline materials as well as metals and oxides. In oxides, they are often charged, which can lead to significant changes in light transmission [11]. Of most importance are defects found in oxides manifested as oxygen vacancies, which are naturally present in the form of Shottky or Frenkel defects [8]. Point defects play a significant role in determining the physical and chemical properties of inorganic materials. This holds for both the bulk and surface properties of oxides, where several kinds of point defects exist and exhibit complex chemistry. Depending on the electronic structure of the material, the nature of oxygen vacancies changes dramatically.

The metal oxidation state in oxides can vary widely. The great variety of behaviors implies that the geometric and electronic nature of oxygen vacancies also change dramatically from case to case. Oxygen vacancies are known as color centers, or F centers (from *Farbe*, the German word for color); if the vacancy is localized at the surface a subscript "s" is added, F_s. F centers exist in three states depending on their electronic charge. Diamagnetic oxygen vacancies can be either neutral (F) or doubly charged (F^{2+}). In the former case, two electrons are associated with the vacancy. Of particular interest for the characterization of oxygen vacancies are the paramagnetic F^+ centers, consisting of a single electron trapped in the cavity formed by removing an oxygen anion, O^- [8].

The influence of γ-radiation onto different types of thin films has been explored. The irradiation of thin films of poly(methyl methacrylate) doped with spirobenzopyran resulted in a permanent change in the materials properties from a nonfluorescent form to a fluorescent, under the excitation of the wavelengths of 488 and 514 nm [12]. Arsenic sulfur (AChS) thin films deposited by thermo-vacuum method were used as radiation-sensitive elements [13]. They were characterized by irreversible changes in their optical properties. Such dosimeters operate by the combined effect of photo-radiation in arsenic sulphur associated with defect recharging processes. However, these devices are complicated in practice and have a limited working range of doses. Throughout this work, a novel approach is used, where oxides in the form of thermally evaporated thin films are considered appropriate cost-effective materials for radiation sensing.

4.2.1.2 Metal Oxides Mixtures

Numerous metal oxides and their mixtures in different proportions in the form of thin films were studied in terms of their susceptibility to the exposure of gamma radiation. Mixing different oxides in various proportions that can control the properties of the semiconductor films were understood [14, 15]. The properties of SiO and In_2O_3 mixed thin films fabricated by the coevaporation technique were reported [14]. Films of SiO have an open structure, which contains a large number of dangling bond centers, and the density of these centers decreases as the indium oxide content is increased in the complex SiO/In_2O_3. This gives rise to an increase in the porosity of the resulting film, and consequently the optical energy gap decreases. Simultaneous sputtering of ZnO and In_2O_3 targets resulted a formation of a $ZnO^-In_2O_3$ compound system with $Zn_3In_2O_6$ and $Zn_2In_2O_5$ phases, which differ in their carrier mobility. A statement has been made that the carrier concentration increases with increasing Zn supply and that the Zn atoms take oxygen away from In_2O_3 bulk through the formation of ZnO [15].

4.2.1.3 Changes in the Properties of TeO$_2$ Thin Films Caused by Radiation

Optical Properties of TeO$_2$ Thin Films Optical properties of thermally deposited TeO$_2$ thin films having a thickness of 50 nm showed a high sensitivity to gamma radiation [16, 17]. Typical plots of the absorption spectra for as-deposited and γ-irradiated thin films of TeO$_2$ are shown in Figure 4.1. The optical band gap

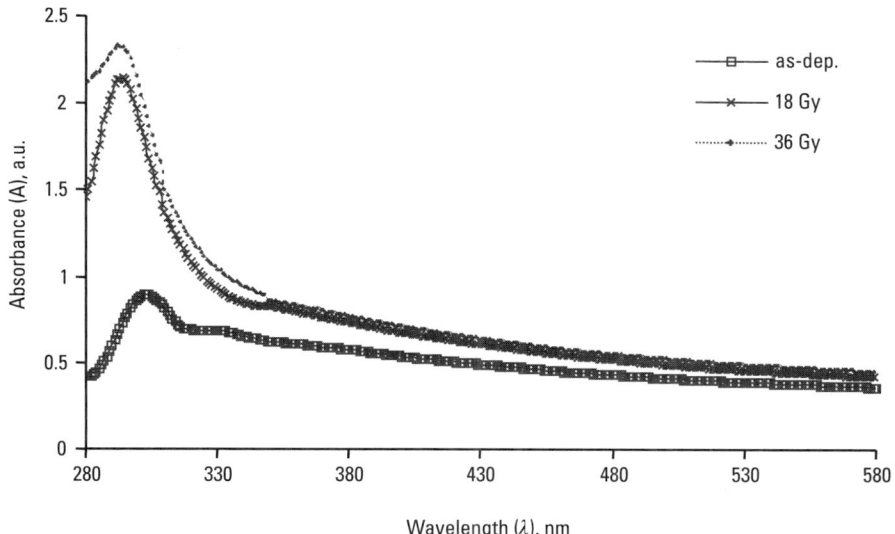

Figure 4.1 The absorption spectra for as-deposited and irradiated TeO$_2$ thin films. (*From*: [17]. © 2003 IEEE. Reprinted with permission.)

value decreased from 3.75 eV to 3.45 eV, with the increase in radiation dose up to 36 Gy. The values of the optical band gap (E_{opt}) for as-deposited and γ-irradiated specimens were estimated using the Mott and Davis model for direct allowed transition [18]. The variation in the optical energy gap with irradiation can be explained using the density-of-state model. It is known that E_{opt} decreases with the increase in the degree of disorder of the amorphous phase [19]. At this stage, one may expect that a band tail would probably be created due to irradiation. The decrease in E_{opt} leads to a shift in the band tail ΔE toward the higher energy region, and hence the calculated value of ΔE is expected to increase as the radiation dose is increased. Generally, wide band gap compound semiconductors, in particular glassy materials, are of interest for room-temperature ionizing radiation detection. However, pure amorphous materials are preferred over typical glasses. In turn, pure crystalline materials are preferred over the pure amorphous because of their greater number of states and their structural characteristics, such as greater susceptibility to permeation [11].

Electrical Properties of TeO₂ Thin Films The values of sheet resistance R_s for as-deposited and gamma-irradiated TeO_2 thin film samples having a thickness of 220 nm were measured with a four-point probe [16]. The data revealed that the sheet resistance linearly decreased from 36.4 kΩ/square to 28.7 kΩ/square with an increase in radiation dose up to a level of 108 Gy. In general, sheet resistance of thin films decreases by the influence of gamma radiation due to the occurrence of more incomplete oxidation and an increase in oxygen vacancies. This is probably due to a change in crystallinity or conduction mechanism.

Radiation-induced changes in material properties have been explored using the Hall effect measurements in [20, 21]. Tolpygo et al. [20] have studied the effect of electron irradiation on critical superconducting temperature T_C, resistivity ρ, and Hall coefficient of thin $YBa_2Cu_3O_{7-\delta}$ films fully loaded with oxygen. The electron irradiation used had energy level below the threshold energy for the formation of in-plane defects. A significant increase in both the slope of resistivity curves and the Hall coefficient have been found to occur due to irradiation. Results on Hall effect and resistivity measurements of low- and high-resistivity silicon, used for fabrication of detectors, nonirradiated, and irradiated with ions (Kr) and neutrons have been presented in [21]. The changes in the structural and the electrical properties were due to the damage caused by irradiation of silicon. An increase in the resistivity of $YBa_2Cu_3O_{7-\delta}$ films with electron radiation was reported [20]. Resistivity measurements in [21] indicate the increase of ρ with increasing fluence Φ for neutrons and for charged particles.

Irradiation with high-energy particles produces electrically active p-type centers. P-type semiconductor materials should lower their resistivity, in accordance with a model, which assumes that lattice vacancies and interstitial atoms are created by irradiation in equal numbers and that each has two energy levels associated with it [22]. The levels of interstitial atom are widely separated in the energy gap: one near the conduction band edge and one near the valence band edge. Both are filled with an electron if the interstitial atom is electrically neutral. The lattice vacancy also has two associated levels, which are close together below the center of the energy gap and are empty for electrical neutrality. In thermal equilibrium at room temperature for n-type material of moderate resistivity (the Fermi level near the center of the gap), the three lower levels will be filled and the upper one is vacant, giving rise to singly ionized interstitials and doubly ionized vacancies. If the material is of moderate p-type resistivity, one of the electrons required to fill the levels of each vacancy will come from the upper interstitial level and the other from valence band. Taking electrons from the valence band decreases the resistivity of a p-type sample by increasing the number of holes [22].

While irradiation usually leads to the creation of structural defects, the healing effect of irradiation is also known [23, 24]. Metals made of powders preliminarily irradiated by electrons or gamma-irradiation are characterized by the absence of big pores and fine homogeneous grain structure. The average sizes of grains are four to five times lower than that in conventional technology [24]. Ionizing radiation has been found to be widely applicable in modifying the structure and properties of polymers and can be used to tailor the performance of either bulk materials or surfaces [23]. Such a healing occurs due to radiation-stimulated recombination of intrinsic (preexisting) defects or to radiation-stimulated ordering of initially disordered phase. This is known as the small-dose effect, because it occurs at low doses when the concentration of the induced defects does not exceed the concentration of intrinsic defects [20].

The typical change in the values of sheet density ρ_s of charge carriers with the increase in radiation dose for TeO_2 thin film specimens is shown in Figure 4.2 [16]. Since TeO_2 is a p-type semiconductor material, conduction in TeO_2 films is due to holes, which are the majority carriers. This decrease in the majority charges can be attributed to the generation of vacancies under irradiation and charge capture on the defects.

Under the rudimentary model of the charge transfer, the Fermi level in the irradiated structure should shift upward with respect to its position in the nonirradiated structure, filling the holes states in plane bands [25]. For TeO_2 thin films, shifting the Fermi energy level under the influence of radiation has resulted in a decrease in the optical band gap E_{opt} value. In turn, filling the holes states could give an explanation to the observed decrease in sheet density ρ_s of charge carriers with the increase in radiation dose. This approach clarifies the observed correlation between changes in both the electrical and the optical properties caused by radiation.

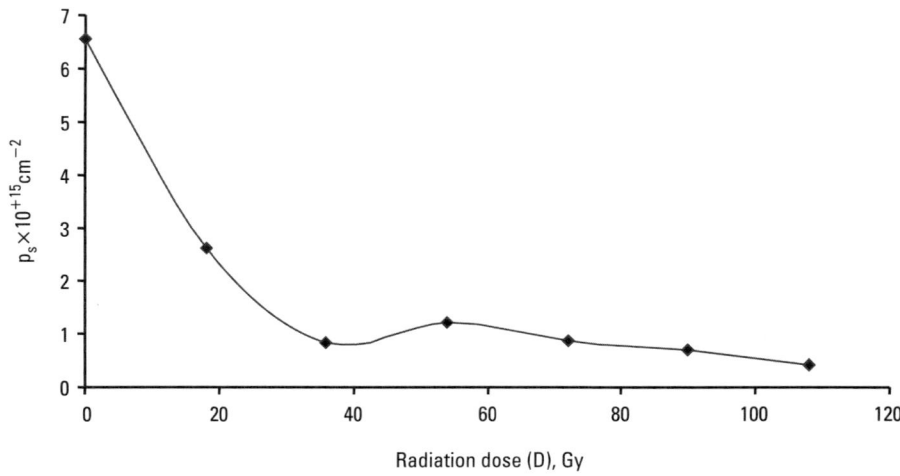

Figure 4.2 Change in the sheet density p_s of charge carriers with radiation for TeO_2 thin films. (*From:* [16]. © 2002 Emerald. Reprinted with permission.)

Oxide Mixtures Thin film samples with 90 wt. percent of TeO_2 and 10 wt. percent of In_2O_3 appeared to be more stable and sensitive to γ-rays in comparison with pure TeO_2 thin films and displayed a monotonic tenfold increase in the values of current with dose up to 90 Gy [17]. The dose response of TeO_2/In_2O_3 thin films was well pronounced, which confirms the advantage of oxide mixing for radiation sensor development.

Also, the properties of a MnO and TeO_2 mixture in the form of thin films were explored in terms of gamma radiation sensing using ^{137}Cs source [26]. Both materials were chosen due to their relatively low melting point, which enables their uniform deposition. To estimate the effect of the dimension on the device sensitivity, the films were chosen to vary in thicknesses of 60 nm and 120 nm. The values of current were observed to increase with the increase in radiation dose up to a level of 684 μGy for MnO films and 1,305 μGy for MnO/TeO_2 films, respectively. Devices with MnO/TeO_2 were found to sustain higher radiation doses and demonstrated a monotonic behavior, which is preferable for personal dosimetry applications. However, the response of the MnO thin films to external exposure was more pronounced. Devices with MnO layers having thickness of 120 nm showed an increase in their values of current with increasing radiation dose up to 850 μGy. While MnO/TeO_2 layers withstand a dose of 1,400 μGy, beyond this dose a decline in the values of current was recorded [26]. In general, thicker films were found to be less sensitive, but they sustained a higher level of dose, confirming the statement given in [27]. To compare the performance of these devices, see Figures 4.3 and 4.4 [26]. They display the dependencies of normalized current $(I-I_0)/I_0$ versus γ-dose under an applied

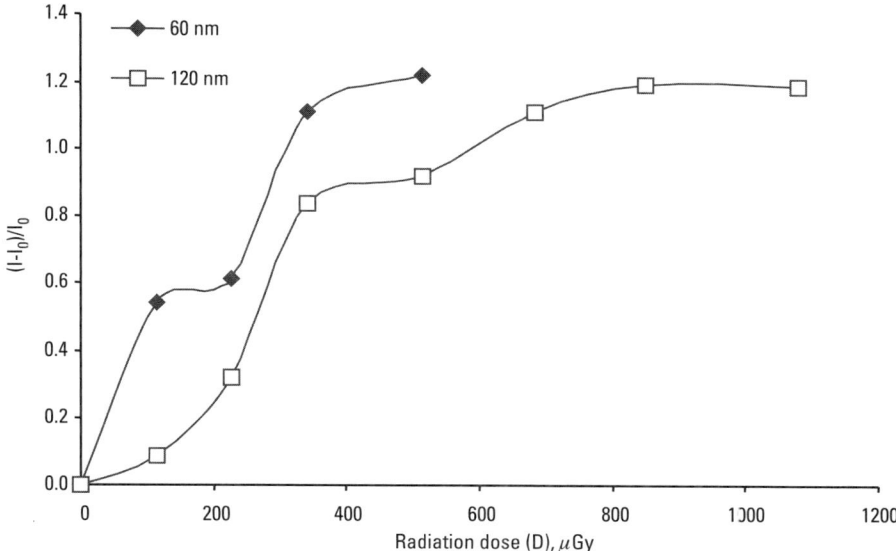

Figure 4.3 Dependence of current $(I-I_0)/I_0$ with dose under an applied voltage of 3V for MnO films having thicknesses of 60 nm and 120 nm. (*From:* [26]. © 2004 Wiley-VCH. Reprinted with permission.)

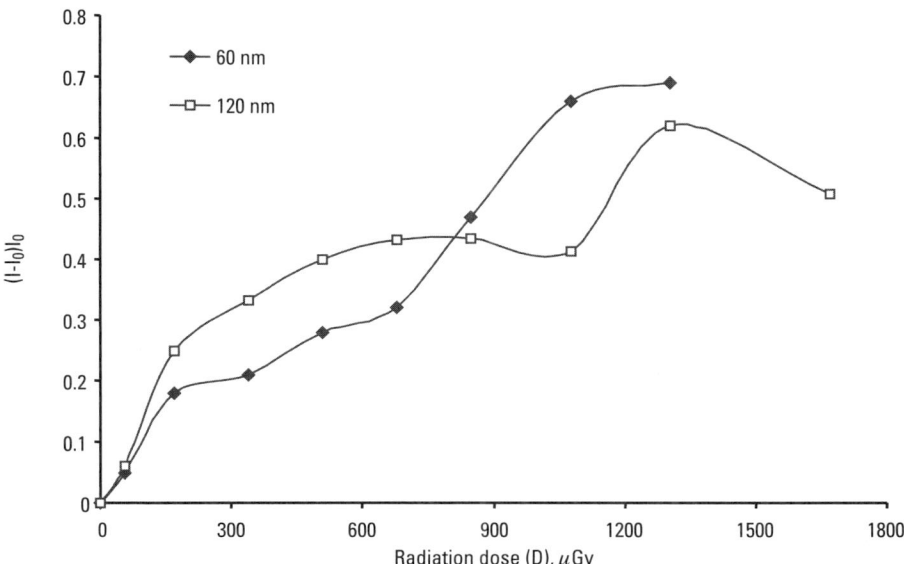

Figure 4.4 Dependence of current $(I-I_0)/I_0$ with dose under an applied voltage of 3V for MnO/TeO$_2$ films having thicknesses of 60 nm and 120 nm. (*From:* [26]. © 2004 Wiley-VCH. Reprinted with permission.)

voltage of 3V for MnO and MnO/TeO$_2$ films, respectively. Sensors made with only MnO exhibited a higher value of normalized current and sustained a lower radiation dose of 850 μGy, whereas the counterpart MnO/TeO$_2$ samples were damaged after a dose of 1,400 μGy. All devices showed an increase in the values of current with increasing radiation dose up to a certain level, which was found to be highly dependent on film thickness and composition [26].

4.2.1.4 Structural Studies

The structural order of a thin film is determined largely by the surface mobility of the adsorbed atoms. A highly disordered amorphous-like structure would be produced when the mobility is negligible, so that the atoms condense at or near the point of impingement. Since oxides have considerably lower atomic mobility than the metals at any temperature, no significant crystallization is expected to take place. However, the nature of thin films could be amorphous or polycrystalline at its best, depending on the materials used and the method of preparation. For example, as-deposited or as-grown Ta$_2$O$_5$ has amorphous structure due to low temperature fabrication, and the films may crystallize during the high-temperature post-formation processes [27]. It should be kept in mind that in order to detect crystallinity with XRD method, a certain size of crystallites is required. For XRD, the lower limit of detection is approximately a volume fraction of the crystalline phase of 0.001. This is why XRD data have to be considered only in terms of the sensitivity of this method.

The influence of gamma radiation on In$_2$O$_3$/SiO films resulted in significant changes in the microstructure of these films. Some kind of agglomerations with variable sizes in the range 0.5–3 μm occurred. After a dose of 8,160 μGy, evidence of partial crystallization was observed with the use of XRD, which is shown in Figure 4.5 [28]. As one may see, the morphological structure of the thin films was highly affected by radiation. Most of the radiation-induced defects are due to oxygen vacancies that to a certain extent cause degradation in the stoichiometry of the layers.

Crystal structure and optical properties of TeO$_2$ were the subjects of numerous theoretical and experimental studies [29, 30]. Tellurium dioxide belongs to the category of compounds in which all the atoms are the so-called p-elements, having nonbonding valence electron pairs. TeO$_2$ and almost all the Te^{IV} containing compounds (including the glasses) exhibit remarkable properties related to macroscopic polarization and polarizability (dielectric, piezoelectric, optic, electro-acoustic), which are of great interest for fundamental science and technology. The origin of such properties, particularly for TeO$_2$-based materials, is related to the peculiarities of the electron distribution inside the coordination polyhedra. At ambient conditions, TeO$_2$ is known to exist in the two polymorphous forms, paratellurite, α-TeO$_2$ (D_4^4 P4$_1$2$_1$2), and tellurite, β-TeO$_2$ (D_{2h}^{15}, Pbca). In both structures, tellurium atoms have four neighboring

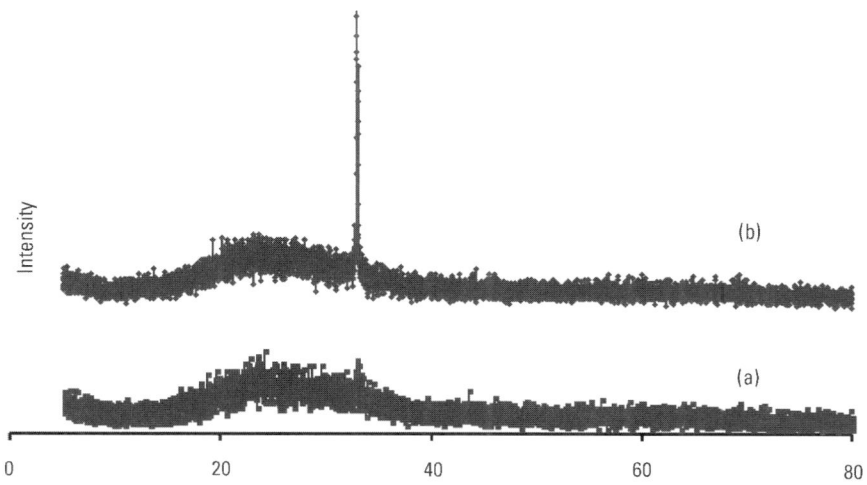

Figure 4.5 XRD pattern for (a) as-deposited and (b) γ-irradiated In$_2$O$_3$/SiO films. (*From:* [28]. © 2005 ttp. Reprinted with permission.)

oxygen atoms, and the basic structural unit is a TeO$_4$ disphenoid, or if the $5s^2$ lone pair of tellurium atoms (E) is taken into account, a distorted TeO$_4$E bipyramid. In this bipyramid, the two equatorial oxygen atoms are separated from Te by distances shorter than the sum of the covalent radii of O (0.73Å) and of Te (1.35Å), and the two axial oxygen atoms by distances longer than that value. Figure 4.6 shows the Raman spectrum of as-deposited TeO$_2$ film in the wavenumber range of 50–750 cm^{-1} [28]. This spectrum could be considered a superposition of Raman spectra of β- and γ-modifications of TeO$_2$ crystals reported by Mirgorodsky et al. [30]. Two strong peaks at 122.78 cm^{-1} and 145.83 cm^{-1} belong to the domain of γ-TeO$_2$ modification, while a peak around 278 cm^{-1} can be assigned to β-TeO$_2$ modification.

After the exposure to γ-radiation, TeO$_2$ film experienced structural alterations, as shown in Figure 4.7. A strong peak appeared at 448.83 cm^{-1}, indicating further transformation to γ-TeO$_2$ modification. Two relatively strong bands were observed for the TeO$_2$-based glass system around 600 cm^{-1} and 700 cm^{-1}. The first is assigned to stretching-mode vibrations of TeO$_4$ bipyramids characteristics of TeO$_2$, while the latter is assigned to stretching-mode vibrations of the TeO$_3$ group, created during the chain-breaking process [28].

4.2.1.5 In$_2$O$_3$/SiO Thin Films as Gamma Radiation Sensors

Optical Properties Figure 4.8 illustrates the dependences of the optical density on radiation dose that were measured at a constant wavelength of $\lambda = 420$ nm for films with two different compositions [31]. Films with composition 50 wt. percent of In$_2$O$_3$ and 50 wt. percent of SiO showed an increase in the values of

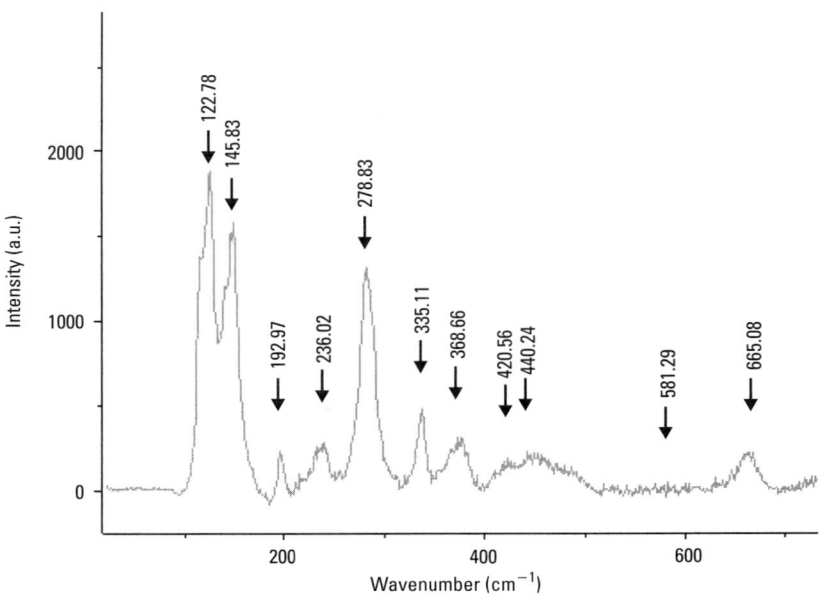

Figure 4.6 Raman spectrum of as-deposited TeO_2 thin film. (*From:* [28]. © 2005 ttp. Reprinted with permission.)

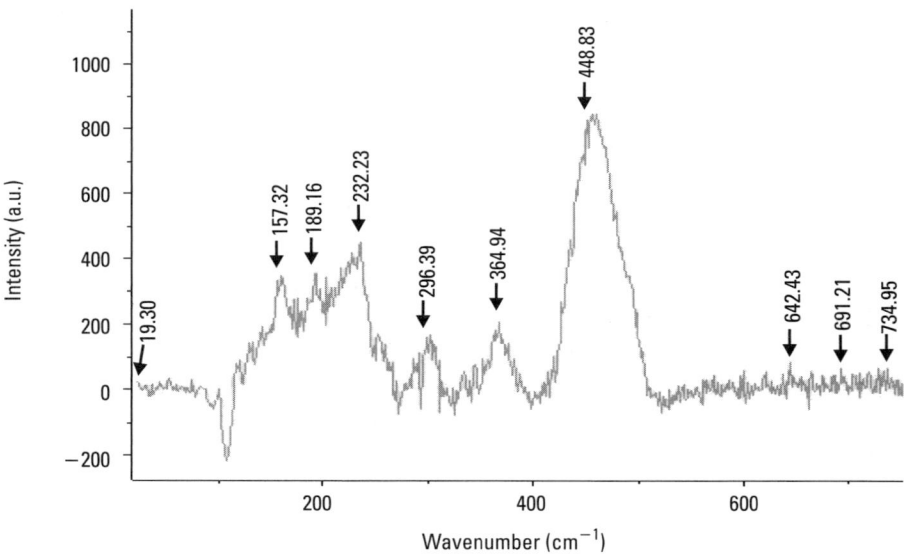

Figure 4.7 Raman spectrum of γ-irradiated TeO_2 thin film. (*From:* [28]. © 2005 ttp. Reprinted with permission.)

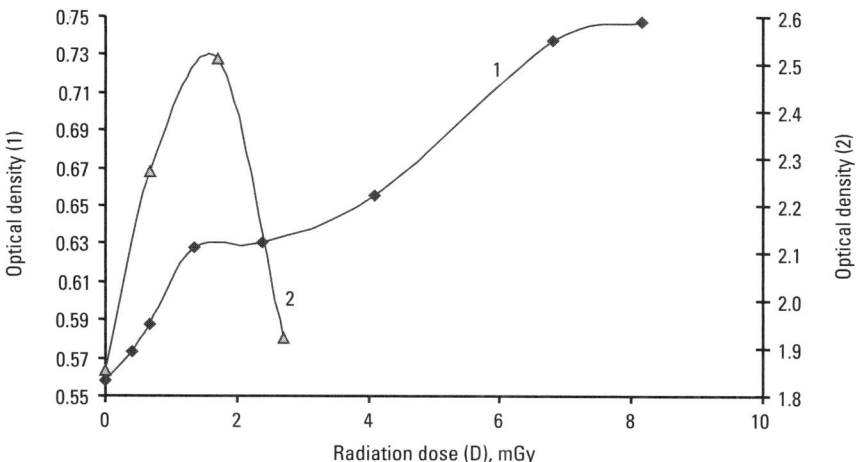

Figure 4.8 Dependences of the optical density on radiation dose at a wavelength of $\lambda = 420$ nm for the films made with (1) 50 wt. percent of In_2O_3 and 50 wt. percent of SiO and (2) with 75 wt. percent of In_2O_3 and 25 wt. percent of SiO. (*From:* [31]. © 1998 SPIE. Reprinted with permission.)

their optical density up to a dose level of 8.16 mGy. Further increase in radiation exposure level resulted in a slow decrease in optical density of these films. The counterpart samples made with 75 wt. percent of In_2O_3 and 25 wt. percent of SiO exhibited a well-pronounced increase in optical density values from 1.852 before irradiation to 2.512 after irradiation with a dose level of 1.7 mGy.

The optical energy gaps E_{opt} for as-deposited and γ-irradiated films were determined from the high-absorption regions of the fundamental edges using the Mott and Davis model assuming indirect allowed transition. Films with 75 wt. percent of In_2O_3 and 25 wt. percent of SiO were very sensitive to low doses of radiation range and showed a strong decline in the optical band gap value from 0.9 eV before irradiation to 0.55 eV after a dose of 1.7 mGy. However, an increase in a dose of up to 2.72 mGy resulted an increase in E_{opt} to 0.83 eV. In contrast, the counterpart films made with 50 wt. percent of In_2O_3 and 50 wt. percent of SiO were less susceptible to incident radiation. They showed a decrease in E_{opt} from 1.15 eV before irradiation to 0.82 eV after an exposure to a dose of 8.16 mGy [31].

The band structure and the existence of an energy gap are believed to be dependent upon the arrangement of the nearest atomic neighbors and the existence of local or short-range order. From the density-of-state model, it is known that E_{opt} decreases with the increase in the degree of disorder in the amorphous phase [19]. The lack of crystalline long-range order in amorphous/glassy materials is associated with a tailing of the density of states into the normally forbidden energy band. The exponential absorption tails (referred to as Urbach's energy,

ΔE) depend on temperature-induced disorder, static disorder, strong ionic bonds, and average phonon energies. Radiation may increase the bond angle distortion so that the optical absorption edge is shifted to the lower energies.

4.2.1.5 Thin Film P-N Junctions

TeO_2/S and $(TeO_2+In_2O_3)/S$ P-N Junctions To investigate the effect of the material composition on device performance, two types of thin film p-n junctions with different p-type sides were manufactured [32]. The first type was pure TeO_2, and the second type was a mixture of 90% of TeO_2 and 10% of In_2O_3. The devices showed an enormous difference in their electrical properties. Figure 4.9 shows plots of typical current-voltage characteristics for two types of samples, where type 1 corresponds to TeO_2/S and type 2 corresponds to $(TeO_2 + In_2O_3)/S$ p-n junctions. The current-voltage characteristic of a sample with $(TeO_2 + In_2O_3)/S$ structure is shifted to the left-hand side of the counterpart sample with TeO_2/S p-n junction. This is caused by the difference in the level of doping of p-type material used in the two types of the p-n junctions. The samples with the mixture of In_2O_3 and TeO_2 sustained higher breakdown voltage.

The irradiation of TeO_2/S diodes assembled on a glass substrate with a ^{60}Co gamma source at a dose rate of 6 Gy/min to a level of 72 Gy has led to a linear increase in the value of leakage current, as can be seen in Figure 4.10. As one may consider the fabricated p-n junction a simple model of radiation sensor, this dependence could serve as a reference to receive information about the radiation dose absorbed [32].

Thin Film P-N Junctions on Silicon Wafer When external excitation, such as light or radiation, interacts with the semiconductor material, it raises a number of electrons from the valence band to the conduction band, creating an equal number of free holes and electrons. Since the temperature is not increased, the energy width of the Fermi occupation probability is the same as it was before the external excitation. But now this is quasi-Fermi probability function for electrons and holes, respectively [22]. A greater intensity of light or radiation will cause the two quasi-Fermi levels to separate farther. The greater external excitation will increase the percentage occupancy of the states of both bands and require adjustments of both quasi-Fermi levels. The region of the band gap between the two quasi-Fermi levels is no longer in thermal equilibrium with the enhanced free-carrier densities. Any impurity energy states laying in this region can act as trapping centers for the recombination of holes and electrons.

External excitation generated excess densities of minority electrons in the p-type region and minority holes in the n-type region. These excited densities are able to influence the population of any deep trapping levels throughout the excited volume near the p-n junction, including the space charge region

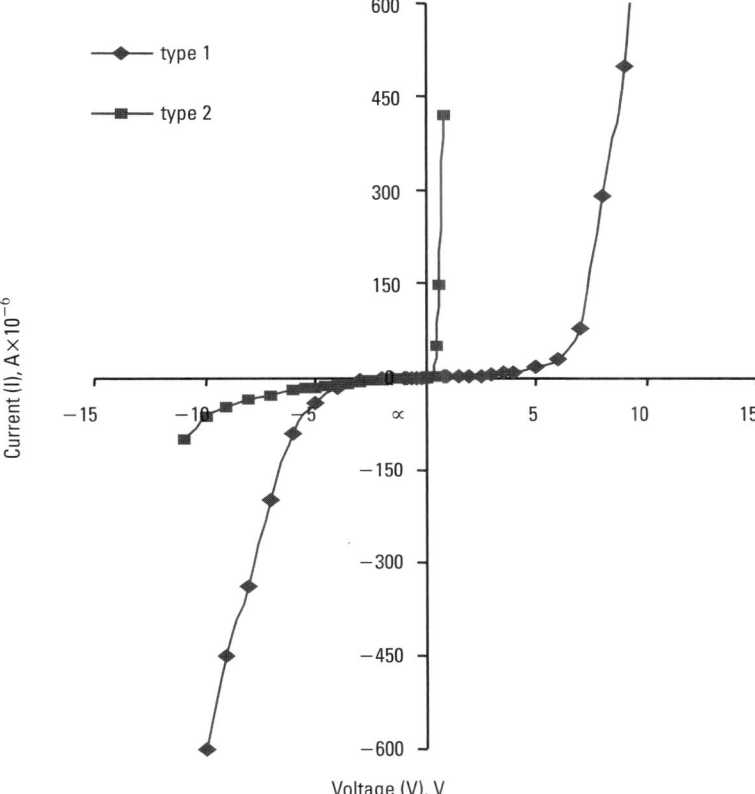

Figure 4.9 Plots of typical I-V characteristics for two types of samples, where type 1 corresponds to TeO$_2$/S and type 2 corresponds to (TeO$_2$ + In$_2$O$_3$)/S p-n junctions. (*From:* [32]. © 2004 Elsevier. Reprinted with permission.)

itself [33]. On removal of the perturbing influence, the return from this excited state towards equilibrium.

A number of p-n junctions were manufactured on Si wafers. Figure 4.11 shows typical current-voltage plots of TeO$_2$/Si samples as-deposited and γ-irradiated with a dose of 342 μSv [32]. As-deposited TeO$_2$/Si structures exhibited a strong p-n junction behavior.

Figure 4.12 shows plots of log (I) versus voltage (V) for TeO$_2$/Si samples: as-deposited and γ-irradiated with a dose of 342 μSv [34]. For the given material combination, changes in an ideality factor η from 2 (recombination in the space charge layer) to 1 (minority carrier diffusion) with the increase in the value of current indicate the importance of both processes.

Leo Esaki first reported this type of diode behavior for p-n junctions, formed between very heavily doped P and N regions due to the tunneling

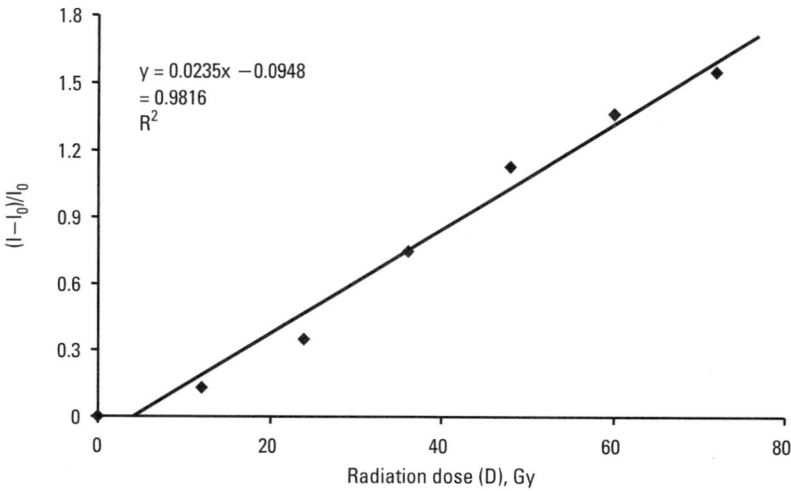

Figure 4.10 Dependence of normalized current $(I-I_0)/I_0$ with radiation dose under an applied voltage of −6V for Al-TeO$_2$/S-Al thin film p-n junctions. (*From:* [32]. © 2004 Elsevier. Reprinted with permission.)

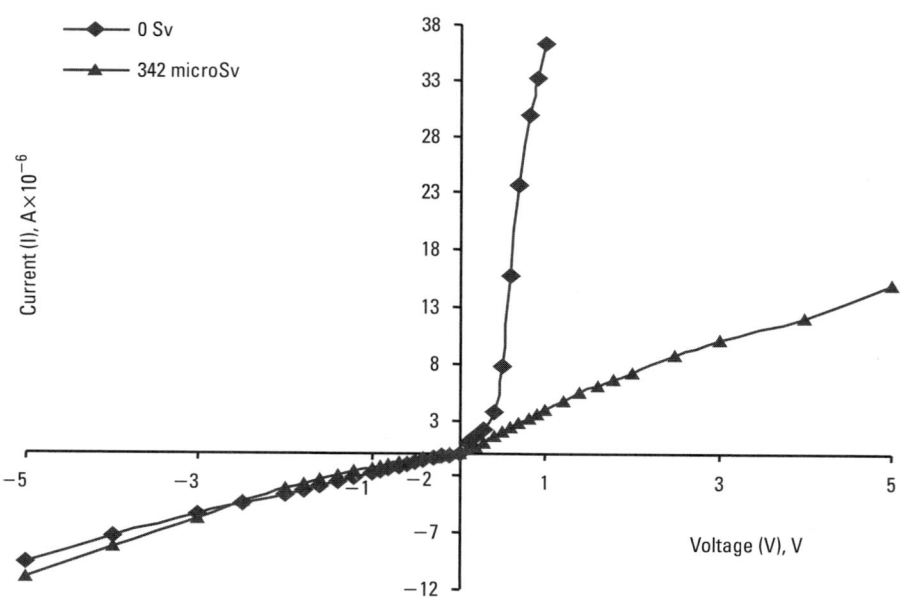

Figure 4.11 I-V plots of as-deposited and γ-irradiated with a dose of 342 μSv TeO$_2$/Si samples. (*From:* [32]. © 2004 Elsevier. Reprinted with permission.)

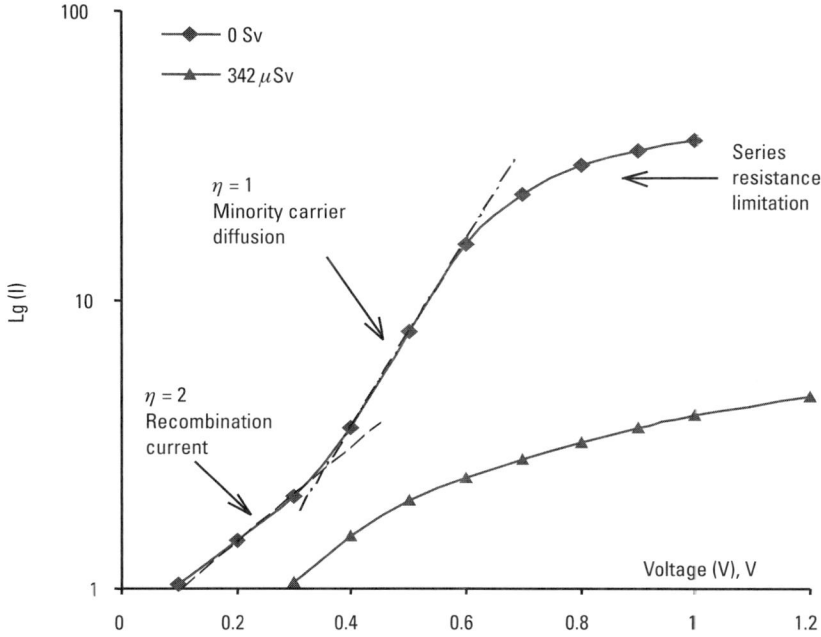

Figure 4.12 Lg (I) versus V plots for TeO$_2$/Si samples: as-deposited and γ-irradiated with a dose of 342 μSv. (*From:* [34]. © 2003 IEE. Reprinted with permission.)

phenomenon [35]. Heavy doping on both sides of the p-n junction reduces the thickness of its depletion region. It also tends to bring the bottom of the conduction band on the n-type side down into line with the top of the valence band in the p-region. In case of the Esaki tunnel diode, the Fermi level must be located inside one or both of the allowed bands. A reverse bias causes the bands to overlap and the barrier to thin down, so that a large tunneling current will flow. A forward bias produces the normal forward characteristic of a p-n junction. Such diodes have a much sharper resistance nonlinearity than the normal diode [22]. Backward diodes are widely used for rectification of small signals in microwave detection, signal mixing, and for very fast switching in high-frequency applications, as they show no appreciable charge-storage effects.

Figure 4.13 illustrates the dependence of normalized current $(I-I_0)/I_0$ with dose under the applied voltage of -1.8V for (In$_2$O$_3$ + SiO)/Si backward diodes [32]. A deterioration of the leakage current is usually observed when the layers are crystallized. The current increase was caused by the grain boundaries, which serve as current paths resulting in poor leakage current characteristics. Another reason for the worsening of the electrical properties is that the films are damaged by the creation of radiation defects in the form of broken Si-O and/or In-O bonds. It is reasonable to assume that the increase in the leakage current after the

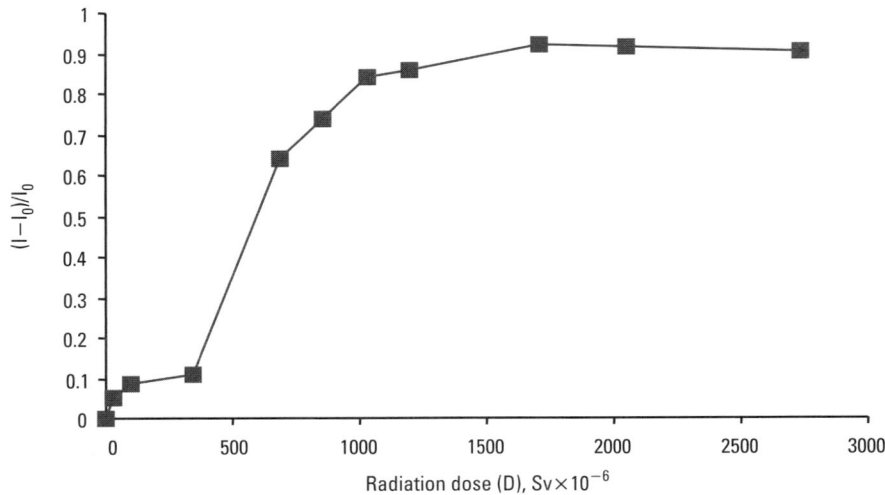

Figure 4.13 Dependence of normalized current $(I-I_0)/I_0$ with dose at $-1.8V$ for $(In_2O_3 + SiO)/Si$ backward diodes. (*From:* [34]. © 2003 IEE. Reprinted with permission.)

influence of γ-radiation is partially attributed to the lowering of the barriers' height at Al/TeO$_2$ and Al/(In$_2$O$_3$+SiO) interfaces. Similar effect of γ-rays was observed in the Al/Ta$_2$O$_5$ interface due to building up a charge in Ta$_2$O$_5$ near this contact [27]. During irradiation process, modification of Si-TeO$_2$ and Si-(In$_2$O$_3$+SiO) interface takes place as a result of oxidation of Si wafer, leading to the enlargement of the mixed transition region, where SiO$_2$ and the intermediate oxidation states of Si coexist. The latent defects, which are activated during irradiation, are in the form of oxide traps and are also responsible for deterioration of the device characteristics.

4.2.2 Metal-Substituted Phthalocyanines Thin Films

4.2.2.1 Metal-Substituted Phthalocyanines

In the last 20 years, many phthalocyanine derivatives have become important materials for use in diverse fields, such as laser printers and photocopiers, gas sensors, optical logic displays, solar cells, and color filters [36]. These applications were made possible by the unique properties of phthalocyanines, such as their stability at temperatures below 773K, low sublimation pressure, chemical stability, and resemblance in molecular structure to hemoglobin and chlorophyll. In general, phthalocyanines are p-type semiconductors with a large band gap that can be altered by changing the metal constituent [37]. Metal-substituted phthalocyanines (MePc) in the form of powder or as thin film structures exhibit polymorphic metastable states of α-forms and stable β-forms [38]. The main distinction between the α- and β-forms is in the stacking angle of the molecular plane with respect to the stacking axis. When ionizing radiation imparts its energy in the form of heat, it causes stretching in the molecular plane, resulting in a change in the stacking angle, thus transforming the material to a different polymorphic state.

4.2.2.2 CuPc Thin Films

The effects of γ-radiation on the absorbance bands, permittivity, and conduction mechanisms of thermal vacuum deposited CuPc thin films were investigated [39]. Figure 4.14 demonstrates the ultraviolet-visible (UV-VIS) absorbance spectra for as-deposited and γ-irradiated CuPc thin films. These are believed to result from the interaction of the ultraviolet and visible light with molecular orbitals in the 18π configuration as well as from the overlapping of electronic orbitals on the central copper atom. Both B- and Q-absorbance bands experienced large rises, when CuPc thin films were exposed to γ-rays. Figure 4.15 shows a linear relationship between the absorbance and radiation dose at the center of the Q-band (615 nm) [40].

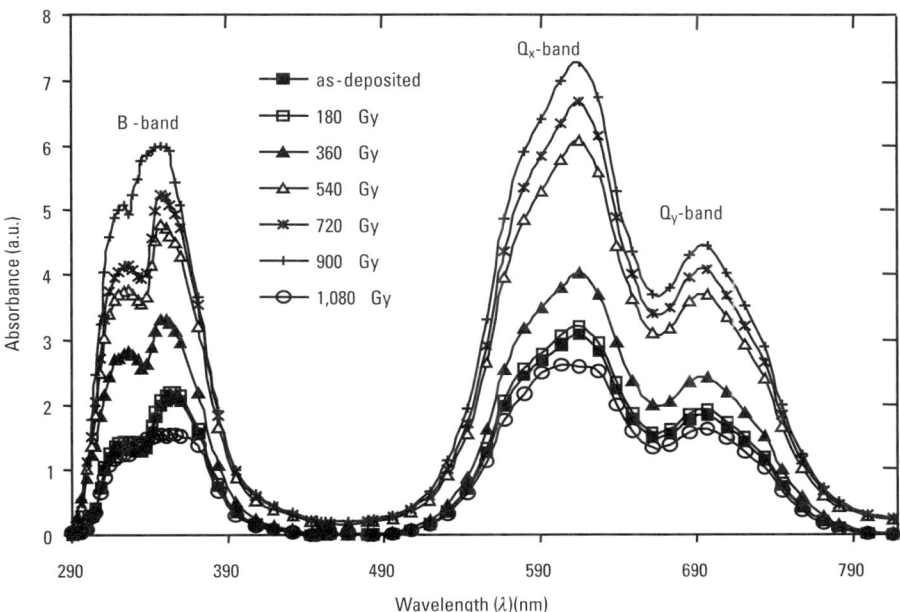

Figure 4.14 UV-VIS absorbance spectra for CuPc thin films illustrating the effects of γ-radiation on B- and Q-absorbance bands. (*From:* [39]. © 2001 IEEE. Reprinted with permission.)

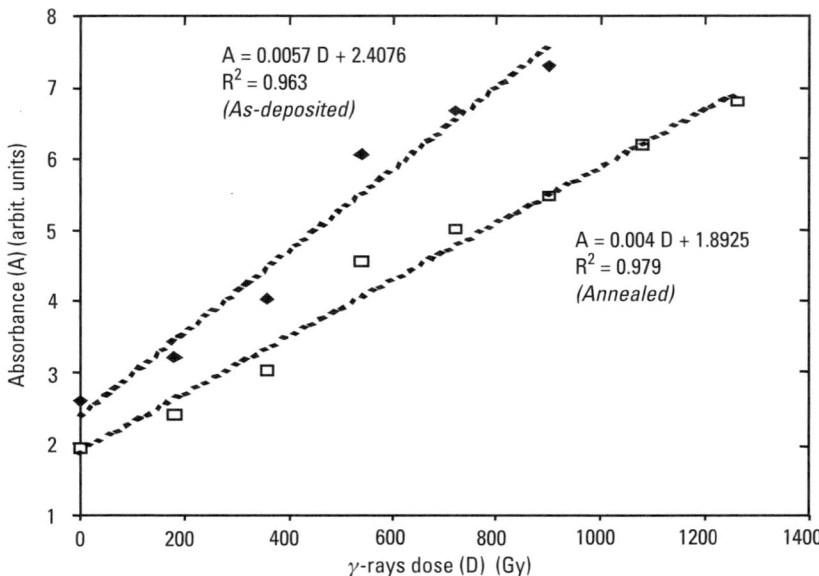

Figure 4.15 Linear dependence of absorbance at the center of the Q-band (615 nm) on γ-ray doses: the annealed sample displayed a higher radiation tolerance than the as-deposited sample. (*From:* [40]. © 2004 IEEE. Reprinted with permission.)

4.2.2.3 Effect of γ-Radiation on the Conduction Mechanisms of CuPc Thin Films

Figure 4.16 shows the various curves of current versus voltage for the as-deposited, annealed, and irradiated CuPc thin films. One can clearly see that I-V relationships for the irradiated samples are different from those of the as-deposited and annealed, implying a change in the conduction mechanism. The I-V characteristic curves of the as-deposited and annealed CuPc films have shown ohmic conduction at low voltages and space-charge-limited conduction (SPCLC) at higher voltages.

Many studies relating to the electrical properties of CuPc thin films have been conducted, and space-charge-limited conduction mechanism manifested itself as the most predominant [41, 42]. However, when the CuPc thin films were irradiated, their I-V characteristics have shown nonpower-law profile, indicating that the SPCLC no longer exists. The change in the conduction mechanism of Al/CuPc/Al thin film devices from a space-charge-limited conduction mechanism to a Poole-Frenkel may be attributed to the increase in carrier concentration, caused by γ-radiation exposure. As the carrier concentration within the CuPc thin film increases, the coulomb potential barrier in the metal-insulator interface rises, leading to higher bulk resistance than the SPCLC. Hence, the current will cease to rise rapidly for $V > 1.8V$ [4]. This is clearly illustrated by

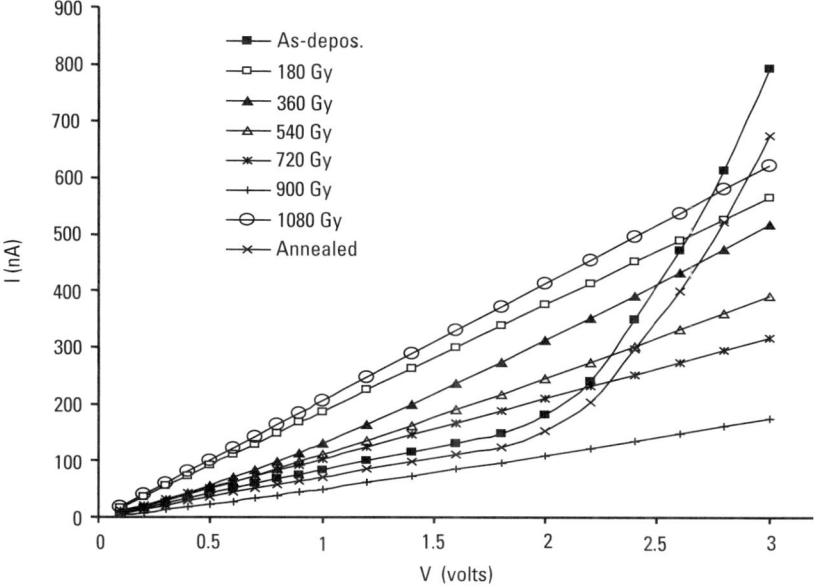

Figure 4.16 I-V characteristics for CuPc thin films. The curves of the as-deposited and annealed samples show a linear ohmic behavior for less than 1.5V and SPCLC for greater values.

I-V characteristics in Figure 4.16, where the apparent space-charge-limited conduction for the as-deposited films is replaced by Poole-Frenkel conduction mechanism for the irradiated films.

4.2.2.4 Effects of γ-Radiation on the Density of Color Centers

The values of the density of color centers can be estimated using Smakula's expression [43]:

$$Nf = 0.89 \cdot 10^{17} \frac{n}{\left(n^2 + 2\right)^2} u \cdot \alpha \qquad (4.1)$$

where N is the charge carrier concentration or the color center density, f is the oscillator strength, which is assumed to be unity, u is the FWHM of the absorption band of the optical spectra, and α is the absorption coefficient at the center of the absorption band.

Figure 4.17 displays a relationship between the density of color centers and γ-ray dose for the as-deposited and annealed Ag/CuPc/Ag thin film structures. The annealed films exhibited better linearity than the as-deposited sample. The experimental data show a dosimetry range of 180–900 Gy for the

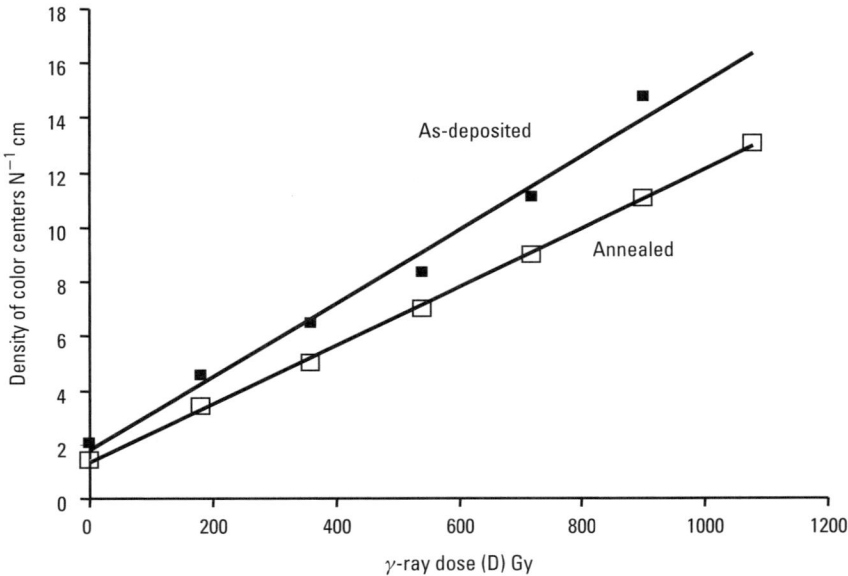

Figure 4.17 Relationship between the density of color centers and γ-ray dose for Ag/CuPc/Ag thin film structure.

as-deposited thin films and 180–1,080 Gy for the annealed samples. Lower density of color centers exhibited by the annealed CuPc films is believed to result from the better-ordered structure. In other words, annealing is known to cause the small islands that originally comprise the as-deposited thin film matrix to fuse together and grow into larger islands or grains. This leads to a reduction of the structural disorders—thus, the density of color centers decreases.

4.2.2.5 Response Characteristics

Usually, radiation sensors have linear dose-response characteristics in certain regions of doses. To cover more than one energy or type of radiation, the approach of using devices with combined structure is widely used [44, 45]. The idea consists of covering parts of the device with filters to absorb some fraction of the incident radiation, thereby modifying the device response such that the ratio of the sensitivity of one filtered area to that of another is a function of the photon energy. The idea of using the filters can be extended: sections of the radiation sensor could differ in material thickness or composition.

The possibility of optimizing the response signal level is also attractive from the electronic circuit design point of view. When a system exhibits clearly pronounced response to external effects such as radiation dose, there is no need for sophisticated circuitry to amplify the output signal. By contrast, if the changes in the output signal are small, it may result in misinterpretation of the received information (e.g., level of absorbed radiation dose). In this case, the experimental error is commensurate with measurement uncertainty. It must be considered that most of the sensors could be susceptible to the environmental conditions, such as temperature, humidity, and electromagnetic field. This may result in under- or overestimation of the absorbed dose. To eliminate such effects, there is a need for corresponding signal conditioning and shielding.

The annealing process is vital for radiation-hardened systems design. Zhu [25] reported that samples annealed under different conditions demonstrated diverse behavior. Samples that were not annealed had the worst radiation hardness. Samples annealed in oxygen were more radiation hard than that annealed in air. Samples annealed under the optimized oxygen conditions were the best in terms of radiation hardness.

4.3 Thick Films as Radiation Sensors

4.3.1 Effects of γ-Rays on the Optical and Electrical Properties of Metal Oxide Thick Films

4.3.1.1 NiO-Based Thick Film Structures

NiO thick film samples having a thickness of 10 mm showed an increase in their optical density from 1.27 to 1.49, with a corresponding increase in the dose to 240 Gy at a constant wavelength of $\lambda = 860$ nm, as shown in Figure 4.18 [46]. The goodness-of-fit value of $R^2 = 0.9728$ shows that the estimated regression line is close to linear. Annealing of the irradiated samples for 12 hours at 388K was found to restore the optical properties of the samples. The observed increase

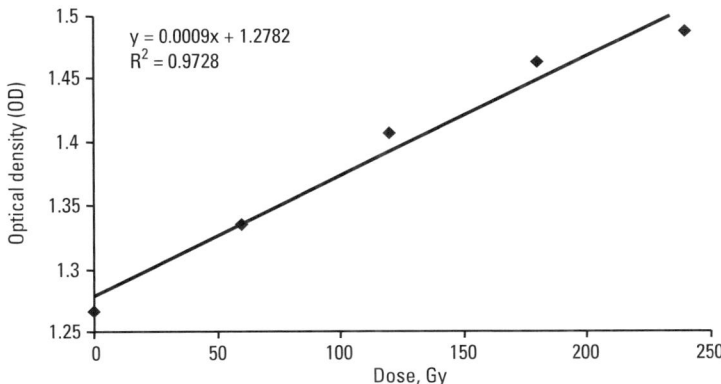

Figure 4.18 Increase in the optical density with increasing exposure dose at a wavelength of $\lambda = 860$ nm for 10-μm NiO thick films. (*From:* [46]. © 2002 IEEE. Reprinted with permission.)

in the absorption coefficient upon annealing in nanocrystalline nickel oxide thin films had been attributed to the structural variations [47].

Planar NiO thick film structures with interdigitated electrodes showed an increase in the values of current for a given applied voltage, with an increase in radiation dose up to 171 μGy [48]. Doping of NiO thick films with 0.1 wt. percent of carbon resulted in larger increments in the values of current and increased the working range to 2,720 μGy. NiO planar structures doped with 0.2 wt. percent of carbon showed an increase in the values of current with an increase in radiation dose up to 1,700 μGy. The substantial signal response obtained came with the tradeoff of lowering the working radiation range of the devices. Therefore, the addition of carbon into the pastes controls the conductivity of the films and their response characteristics, such as sensitivity and dose range.

In general, all devices with interdigitated planar structure showed an increase in the values of current with the increase in radiation dose up to a certain level, which could be considered a working region. The electrical properties of the films were highly dependent on the composition of the oxide materials used. It was experimentally demonstrated that it is possible to fabricate a device that would satisfy the requirement of a particular application, in this case the sensitivity to γ-radiation exposure and working dose region.

The electrical properties of mixtures containing carbon black are widely explained and experimentally supported within the framework of interparticle tunneling conduction and the framework of classical percolation theory [49, 50]. Percolation theory considers, in the simplest case, the formation of clusters in the volume of a matrix. Those clusters, after reaching an appropriate concentration of filler particles, could form an infinite network. In the case of conducting filler particles, the conductivity in the system should increase dramatically after reaching this level, which is named the percolation threshold [50]. Simple models of the ordering of globular particles in the resistor network through the nonconducting matrix show that the conductivity σ of the percolating system depends on the concentration of conducting elements ν as a power law:

$$\sigma = \sigma_0 (\nu - \nu_c)^t \qquad (4.2)$$

where ν_c is the critical volume fraction (percolation threshold), meaning a minimal volume fraction of conducting filler at which a continuous conducting chain of macroscopic length appears in the system. When the concentration of the filler is large (above the percolation threshold), the material exhibits graphitic conductivity, indicating that the conduction network is continuous [51]. The critical exponent t expresses the rate of conductivity change depending on the conducting component concentration, and σ_0 is the conductivity of the conducting phase. It is clear that under these conditions, an increase in conductivity of several orders of magnitude at the percolation threshold results in an extremely high sensitivity of this quantity to the particle content.

A Bruggeman symmetric medium is a composite made up of an infinite range of insulating and conducting particles sizes, as shown here [52]:

$$\phi \frac{\sigma_h - \sigma_m}{\sigma_m + A\sigma_e} + (1-\phi) \frac{\sigma_l - \sigma_e}{\sigma_h + A\sigma_e} = 0 \quad (4.3)$$

where σ_m is the conductivity of the composite, σ_h is the high-conductivity phase, σ_l is the low-conductivity phase, and A depends on the demagnetization coefficient of the spheres building up the media (= 2 for spheres)[53]. The Maxwell-Wagner equations are used to characterize composites with spherical inclusions where the system is made up of an array of coated particles and is given by:

$$\frac{\sigma_m - \sigma_l}{\sigma_m + 2\sigma_l} = \phi \frac{\sigma_h - \sigma_l}{\sigma_h + 2\sigma_l} \quad (4.4)$$

for conducting particles coated with an insulator. The electrical conductivity of a binary system is modeled using the general effective media (GEM) equation given as:

$$\frac{f\left(\sigma_l^{1/t} - \sigma_m^{1/t}\right)}{\sigma_l^{1/t} + A\sigma_m^{1/t}} + \frac{\varphi\left(\sigma_h^{1/t} - \sigma_m^{1/t}\right)}{\sigma_h^{1/t} + A\sigma_m^{1/t}} = 0 \quad (4.5)$$

where $f + \varphi = 1$ [54]. An analytical description of temperature dependence of resistance in a classical two-phase percolation system was proposed for cermet and polymer thick films. Since the effective properties depend on the spatial distribution of the conductive phase in the host medium, there is no universal solution to modeling the electrical conductivity of binary composites. However, the electrical conductivity of carbon black containing composites was found to depend on the structure of the carbon black particles and to deviate from the expectations of classical percolation theory [49]. Generally, a trend can be noticed that the larger the inhomogeneity in carbon black distribution, the smaller is the effective conductivity of the composite. This is why the mixing process requires proper monitoring (e.g., automated shear mixing can be recommended).

4.3.1.2 NiO-Based Thick Film P-N Junctions

Radiation sensing properties of NiO thick films, screenprinted on n-type Si wafers to form p-n junctions were investigated [55]. Carbon doping was used to control the conductivity of NiO thick films. Figure 4.19 shows changes in I-V characteristics of NiO-based p-n junctions with the increase in amount of carbon doping. It caused an alteration in the behavior of the diodes toward a backward diode type.

The values of both the leakage current under the reverse-biased condition and the current when the diodes were forward-biased increased considerably with the increase in radiation dose. The current increase was caused by the grain boundaries, which serve as current paths resulting in poor leakage current characteristics. Another reason for the worsening of the electrical properties is that the films were damaged by the creation of radiation defects in the form of broken Ni-O bonds. During the irradiation process, modification of Si-NiO interface takes place as a result of oxidation of Si wafer, leading to the enlargement of the mixed transition region, where SiO_2 and the intermediate oxidation states of Si coexist. The latent defects, which are activated during irradiation, are in the form of oxide traps and are also responsible for the deterioration of the device characteristics

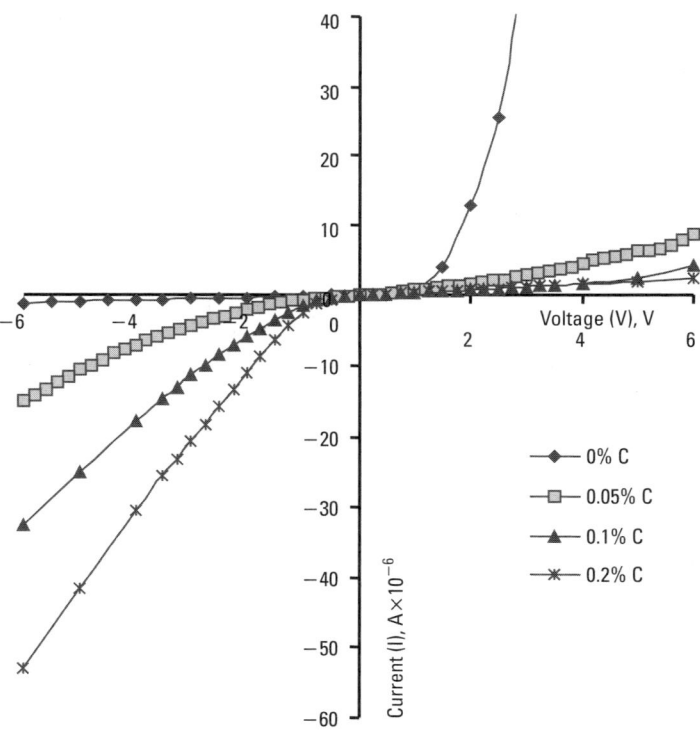

Figure 4.19 I-V plots of NiO/Si structures, doped with various amount of carbon. (*From:* [55]. © 2003 IEEE. Reprinted with permission.)

4.3.1.3 CeO$_2$-Based Thick Film Structures

Cerium oxide (CeO$_2$) is an oxide with fluorite structure, which has the property of deviating from stoichiometry as a function of temperature, pressure, and so forth [56]. Undoped CeO$_2$ is an n-type material, and its conductivity is directly related to oxygen diffusion. Cerium oxide films have received great interest because of their high transparency in the visible and near-IR region and their electro-optical performance. The effect of radiation damage in CeO$_2$ was studied in terms of changes in the oxidation state of Ce and changes in its edge shapes and peak positions [57].

Carbon-doped and pure CeO$_2$ thick films in form of capacitors, resistors, and p-n junctions were explored for γ-radiation dosimetry purpose [58]. Figure 4.20 shows the increased values of current under the applied voltage of 20V for 0.2 wt. percent and 0.5 wt. percent carbon-doped CeO$_2$ resistors. Since these devices exhibited lower resistance compared to undoped CeO$_2$ films, radiation-induced changes in their electrical properties were more pronounced. Resistors doped with 0.2 wt. percent of carbon showed a gradual increase in the values of current with the increase in dose up to 4.42 mSv, whereas resistors with 0.5 wt. percent of carbon sustained a lower dose of 1.7 mSv and were damaged on further exposure. One may notice that films with 0.5 wt. percent of C were much more conductive than the counterpart films with 0.2 wt. percent of C, as they showed values of current in mA and μA, respectively. The substantial signal response of devices with 0.5 wt. percent of C has a tradeoff of lowering the working radiation range of these devices to 1.7 mSv comparatively with 4.42 mSv of the devices with 0.2 wt. percent of C. By changing the structure of the device and the composition of the material used, it is possible to fabricate a

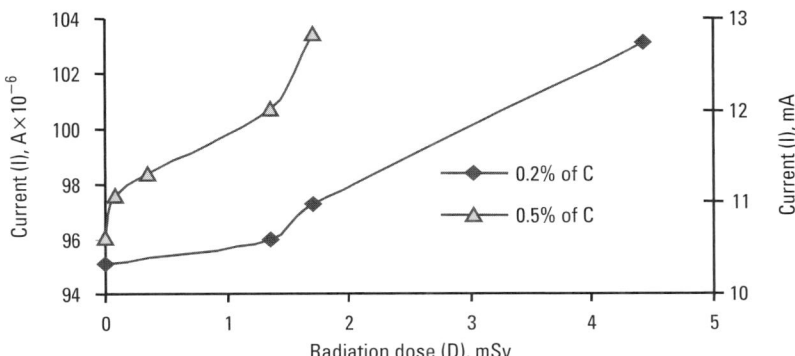

Figure 4.20 Increase in the values of current under the applied voltage of 20V for CeO$_2$ resistors doped with 0.2 wt. percent and 0.5 wt. percent of carbon. (*From:* [58]. © 2004 Elsevier. Reprinted with permission.)

device that would satisfy the requirement of a particular application—in this case the sensitivity to γ-radiation exposure and working dose region.

Figure 4.21 shows current-voltage plots of as-deposited and γ-irradiated CeO_2/Si samples that behaved like backward diodes [58]. The value of leakage current for CeO_2/Si backward diodes increased considerably with the increase in radiation dose up to a level of 570 μSv, while no consistent changes were observed in the values of current when the diode was forward-biased. Figure 4.22 illustrates the dependence of normalized current $(I-I_0)/I_0$ with dose under the applied voltage of $-1.5V$.

Slow initial increase in the values of reverse current was monitored with the increase in radiation to a dose of around 150 μSv. In the second region, from 150 μSv to around 300 μSv, a very fast increase in the values of current was detected. Further increase in radiation dose up to a level of 570 μSv led to some kind of saturation in current-voltage characteristics.

As one compares different cerium oxide thick film structures, CeO_2/Si p-n junction can be found the most sensitive, as they showed considerable increase in the values of leakage current with the increase in radiation dose up to a level of 570 μSv and may be regarded as an alternative to the existing personal radiation sensors. Capacitor-type structures made with undoped CeO_2 thick films sustained a very high dose level of 6.12 mSv and can find their application in aggressive radiation conditions. Irradiation of the samples and subsequent annealing was continuously repeated to ensure the repeatability of the results [58].

4.3.1.4 (In_2O_3 + SiO)/Si Thick Film Sensors

The morphology and structural properties of In_2O_3/SiO thick films with four various compositions were studied [59]. A fine-grained uniform surface of In_2O_3 films with flake-shaped particles measuring 0.2–0.5 μm was revealed with SEM analysis. The surface of the films made with 25 wt. percent of In_2O_3 and 75 wt. percent of SiO possessed large-scale uniformity. The openings between the larger particles of SiO were filled with much smaller particles of In_2O_3, leaving no big pores. The X-ray, diffraction, and Raman spectroscopy data indicated the presence of indium oxide crystal phase in all compositions, whereas none of the silicon oxide–based crystalline phases, such as quartz, stishovite, or cristobalite, were identifiable. Furthermore, the intensity of the In_2O_3 peaks decreases and the intensity of the amorphous halo increases as the amount of silicon oxide is increased. This indicates that SiO must be present in the amorphous state.

Figure 4.23 shows the dependence of the normalized current $(I-I_0)/I_0$ on γ-dose under the applied voltage of $+2V$ for In_2O_3/Si diode [59]. A fast close to linear increase in the value of current under the forward-bias condition is clearly seen at dose of up to 170 μSv.

Diodes made with 75 wt. percent of In_2O_3 and 25 wt. percent of SiO experienced a major rise in the values of current under forward bias up to a dose

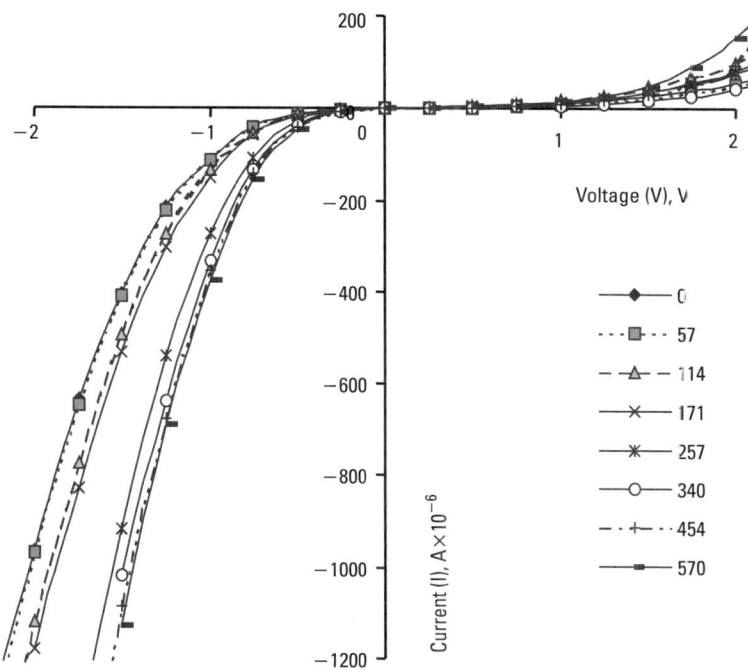

Figure 4.21 Changes in I-V plots of CeO_2/Si backward diode with the increase in γ-radiation dose. (*From:* [58]. © 2004 Elsevier. Reprinted with permission.)

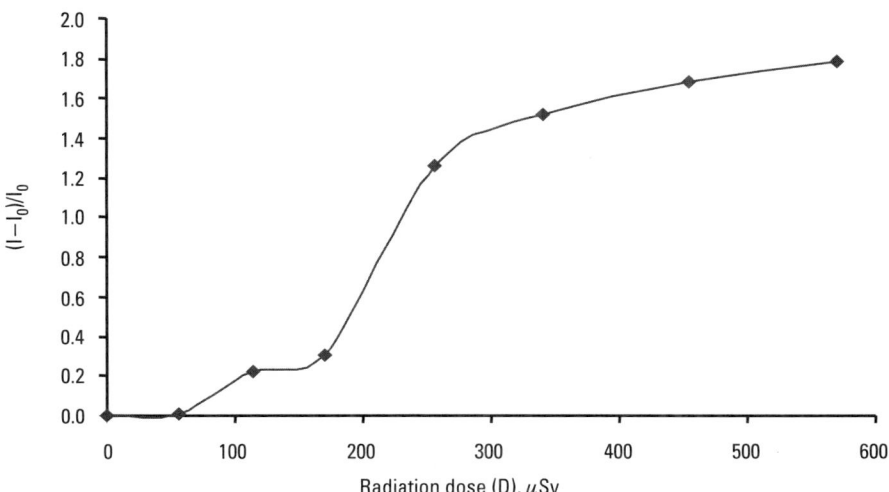

Figure 4.22 Dependence of normalized current $(I-I_0)/I_0$ with radiation dose under the applied voltage of $-1.5V$ for CeO_2/Si backward diode. (*From:* [58]. © 2004 Elsevier. Reprinted with permission.)

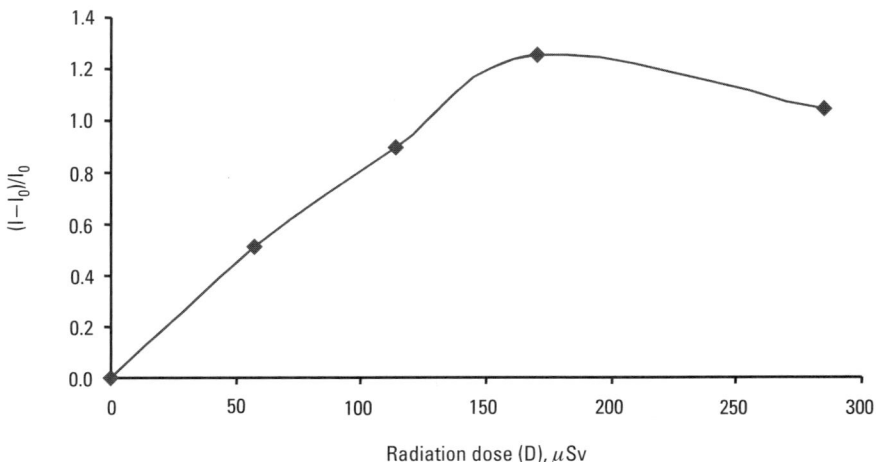

Figure 4.23 Dependence of normalized current (I-I$_0$)/I$_0$ with γ-dose under the applied voltage of +2V for In$_2$O$_3$/Si diode. (*From:* [59]. © 2004 Emerald. Reprinted with permission.)

of 114 μSv. In a range between 114 μSv and 578 μSv, some kind of saturation has occurred with little if any increase in the values of current. After a dose of 578 μSv, a slow decrease in the values of current became noticeable, but in contrast to the diodes with only In$_2$O$_3$, no clear threshold dose was determined after which radiation damage takes place. In a p-n junction made with 50 wt. percent of In$_2$O$_3$ and 50 wt. percent of SiO, the values of current under both forward and reverse biased conditions experienced initial significant increase with the increase in radiation dose up to 114 μSv, while further exposure to a dose of 700 μSv resulted in saturating characteristics. As one can notice, the augmented amount of SiO has led to widening of the radiation dose range that these devices can sustain. After a dose of 700 μSv, a decrease in the values of current took place. Another point that is worth mentioning is the operational voltage scale of the p-n junctions. As the pure In$_2$O$_3$ sample is very conductive, the addition of SiO was found to make the samples more insulative. Figure 4.24 illustrates the dependence of the normalized current (I-I$_0$)/I$_0$ with radiation dose under the applied voltages of -4V and $+4$V for p-n junctions made with 25 wt. percent of In$_2$O$_3$ and 75 wt. percent of SiO.

A number of p-n junctions based on thick films with four different mixtures of In$_2$O$_3$ and SiO components were manufactured and exposed to gamma radiation in identical way, so one can compare the performance of these devices and trace the influence of composition on the response characteristics. Samples made only with In$_2$O$_3$ were the most conductive and gave the most pronounced response to external radiation exposure. However, this makes them the most

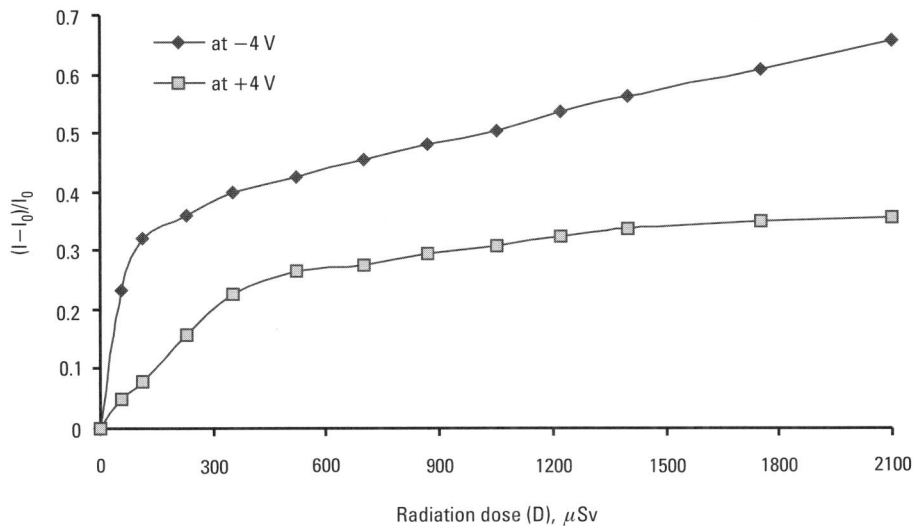

Figure 4.24 Dependence of normalized current $(I-I_0)/I_0$ with γ-dose under the applied voltages of +4V and −4V for a p-n junction made with 25 wt. percent of In_2O_3 and 75 wt. percent of SiO. (*From:* [59]. © 2004 Emerald. Reprinted with permission.)

susceptible to radiation, as they were damaged after a dose of 170 μSv. An increase in the amount of SiO constituent in thick film compositions was found to increase the threshold radiation dose, after which a damage to the diodes can take place. The threshold levels were 578 μSv for films made with 75 wt. percent of In_2O_3 and 25 wt. percent of SiO; 700 μSv and 2,100 μSv for the films made with 50 wt. percent of In_2O_3 and 50 wt. percent of SiO and 25 wt. percent of In_2O_3 and 75 wt. percent of SiO, respectively [59]. Samples made with only pure In_2O_3 are recommended for detection of low levels of radiation. The counterpart samples made with 25 wt. percent of In_2O_3 and 75 wt. percent of SiO are recommended for high-dose application, as they sustained a dose of up to 2,100 μSv.

4.3.2 Effects of γ-Rays on the Optical Density of MePc Thick Films

4.3.2.1 CuPc Thick Films

Six absorbance spectra for CuPc polymer thick films are displayed in Figure 4.25 [60]. It was observed that the increase in γ-radiation dose caused a continuous linear increase in the absorbance and shift of the centers of their B- and Q-bands to higher wavelength in comparison to that of the as-printed films. For instance, when CuPc thick films were exposed to γ-ray doses of 18 kGy, the following changes were monitored [60]:

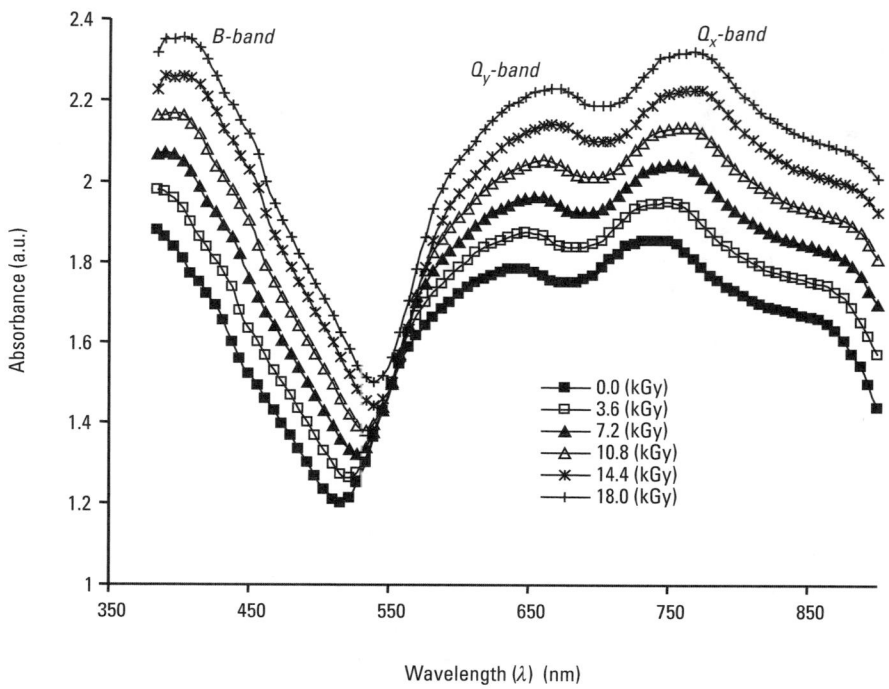

Figure 4.25 Absorbance spectra for the as-printed and irradiated CuPc thick films. (*From:* [60]. © 2002 IEEE. Reprinted with permission.)

1. The center of the B-band shifted from 384 nm to 396 nm, while the absorbance increased from 1.88 to 2.35.

2. The Q_x-band also displayed a similar shift from 738 nm to 762 nm with an increase in the absorbance from 1.85 to 2.32.

3. The center of the Q_y-band shifted from 630 nm to 654 nm with a change in the value of the absorbance from 1.78 to 2.22.

4.3.2.2 NiPc Thick Films

Six absorbance spectra for the NiPc thick films are illustrated in Figure 4.26 [61]. When CuPc thick films were exposed to γ-ray dose of 14.4 kGy, the following changes were observed [61]:

1. The center of the B-band shifted from 354 nm to 420 nm, while absorbance increased from 2.45 to 3.29.

2. The Q_x-band displayed a shift from 678 nm to 732 nm with an increase in absorbance from 2.35 to 3.16.

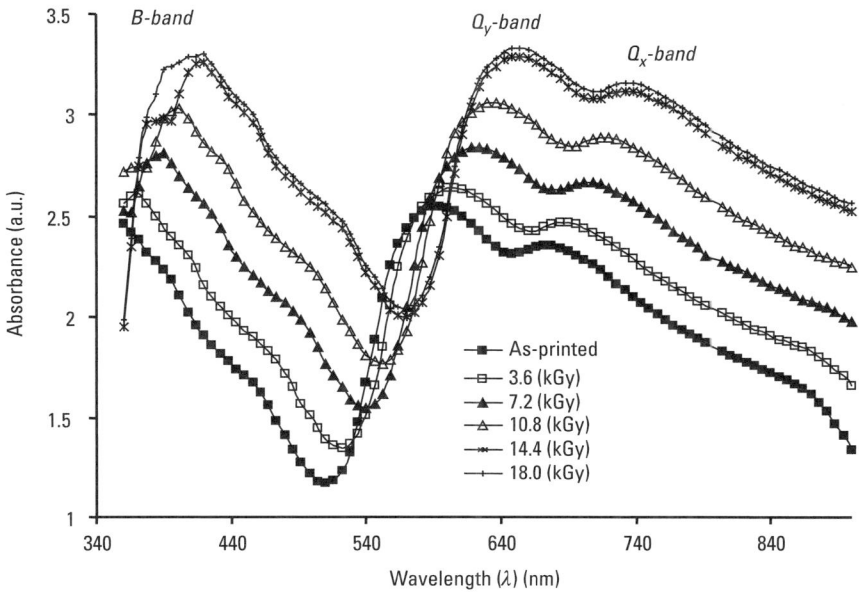

Figure 4.26 Absorbance spectra for the as-printed and irradiated NiPc thick films. (*From:* [61]. © 2002 IMAPS. Reprinted with permission.)

3. The center of the Q_y-band shifted from 594 nm to 648 nm with change in the value of the absorbance from 2.59 to 3.33.

4.3.2.3 MnPc Thick Films

Seven absorbance spectra for MnPc thick films are shown in Figure 4.27 [62]. Similarly to CuPc and NiPc, the MnPc thick films showed an increase in the absorbance and a shift to higher wavelengths of B- and Q-bands centers with radiation. When NiPc thick films were exposed to a γ-dose of 21.6 kGy, the following changes were observed [62]:

1. The center of the B-band shifted from 390 nm to 414 nm, while the absorbance increased from 1.28 to 1.95.
2. The Q-band displayed similar shift from 762 nm to 786 nm with an increase in absorbance from 1.06 to 1.58. The saturation level was reached at γ-rays dose of 21.6 kGy.

4.3.2.4 CoPc Thick Films

Seven absorbance spectra for CoPc thick films are shown in Figure 4.28 [63]. The spectra of cobalt phthalocyanine displayed the lowest optical density, indicating higher structural order compared to other MePc thick films. When CoPc

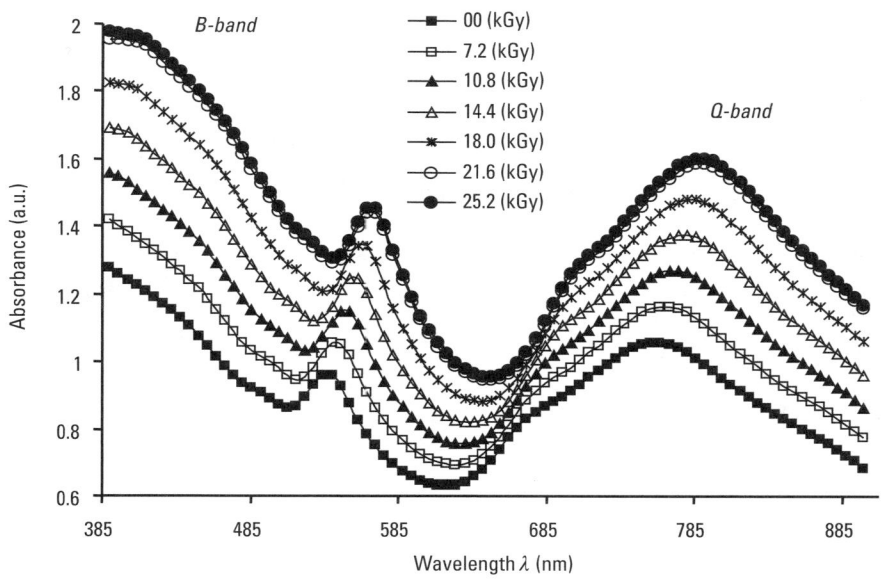

Figure 4.27 Absorbance spectra for the as-printed and irradiated MnPc thick films. (*From:* [62]. © 2002 MDPI. Reprinted with permission.)

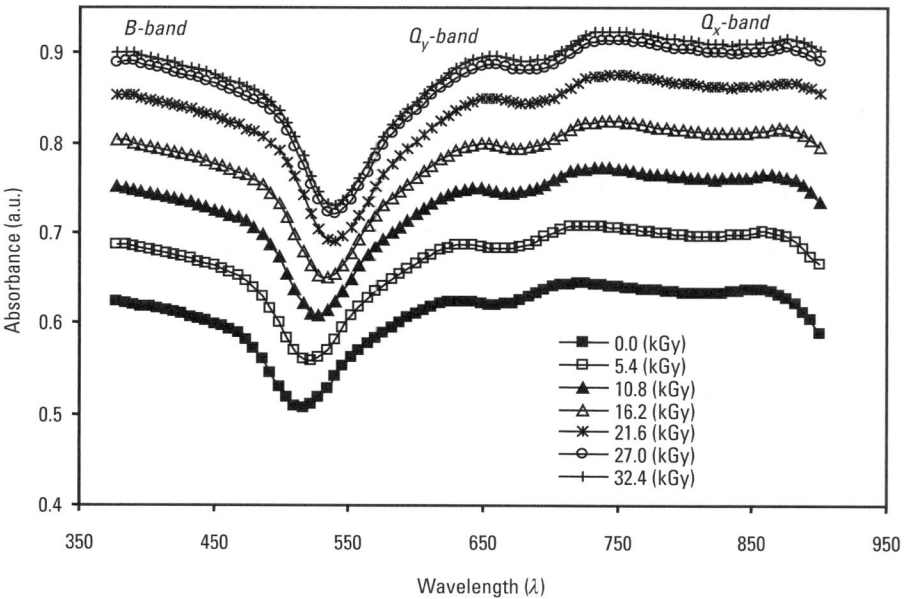

Figure 4.28 Absorbance spectra for the as-printed and irradiated CoPc thick films. (*From:* [63]. © 2002 IMAPS. Reprinted with permission.)

thick films were exposed to γ-ray doses of 21.6 kGy, the following changes were observed [63]:

1. The center of the B-band shifted from 378 nm to 396 nm, while the optical density increased from 0.62 to 0.85.
2. The Q_x-band displayed a shift from 720 nm to 744 nm with an increase in optical density from 0.64 to 0.88.
3. The center of the Q_y-band shifted from 630 nm to 654 nm with a change in the value of the optical density from 0.62 to 0.85. The saturation level for CoPc thick film was 32.4 kGy, which is the highest among all tested MePc films.

The changes in the optical density are attributed to an increase in the density of structural defects that were induced by exposing the thick films to γ-radiation. The analysis of the spectra for all MePc films revealed linear relationships between the optical density and the γ-radiation dose. As an example, Figure 4.29 displays two straight lines, which illustrate the relationship between the optical density and the γ-radiation dose for B- and Q-bands of CoPc thick film [63]. It can be noticed that the optical properties of MePc had different ranges of radiation responses: CuPc thick film displayed a dose range of 3.6–18.0 kGy, NiPc thick film exhibited a dose range of 3.6–14.4 kGy, MnPc thick film demonstrated a dose range of 7.2–25 kGy, and CoPc thick films showed a dose range of 5.4–27.0 kGy. The analysis of the optical absorption

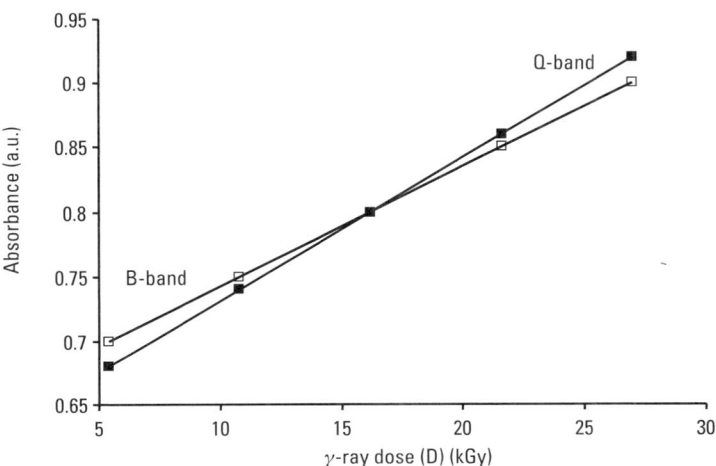

Figure 4.29 Plots of the optical density versus γ-ray dose for B- and Q-band peaks of CoPc thick film. (*From:* [63]. © 2002 IMAPS. Reprinted with permission.)

edge of the CuPc, NiPc, MnPc, and CoPc revealed their optical band gap values of 2.38, 2.30, 1.80, and 2.62 eV, respectively.

4.3.3 MOS Dosimeter Using Bismuth Oxide (Bi_2O_3) and Copper Phthalocyanine (CuPc) Polymer Thick Films

It is generally accepted that silicon dosimeters exploit the generation of charge carriers and holes trapping centers in the gate insulator (SiO_2) of an MOS device, which is caused by the shift in the threshold voltage. This shift is proportional to the exposed radiation dose [64]. The three main factors influencing the sensitivity of any MOS dosimeters are the attenuation of the gate oxide materials to the radiation of interest, the drain to source voltage, and the thickness of the oxide film. The sensitivity of these transistors to radiation is strongly thickness-dependent on the gate oxide. It was indicated that the thickness of the gate oxide employed in dosimetry should be greater than one micron, which is extremely difficult to obtain in case of SiO_2 thin films [64].

As an alternative, other oxide materials having screenprinted thick film structures were explored in order to provide more radiation-sensitive transistors [65]. Polymer thick film of bismuth oxide of 15 μm in thickness was used as a gate material. The bismuth has very large atomic weight (209 atomic units), therefore exhibiting much higher attenuation to γ-rays than the silicon oxide. The applied drain to source voltage is known to exaggerate the susceptibility of the MOS devices to radiation damage, thus increasing their sensitivity [66].

Investigations into the changes in the properties of MOS capacitors is considered to be the simplest and yet the most effective method of understanding the mechanisms of traps formation in the oxide/semiconductor interface under the influence of nuclear radiation. All such mechanisms result in a buildup of a charge density in the oxide layer and a creation of trapping states in the oxide/semiconductor interface. Despite the absence of the threshold channel currents in the MOS capacitors, the entire C-V_G characteristic contains even more information about the oxide semiconductor and their interface states. The flat band and inversion voltages are very important pieces of information: the flat band condition occurs at a point where there is no bending of the copper phthalocyanine conduction band. At this point, the C/C_O has a common value of 0.8 on the depletion side of the C-V_G curve [65]. The voltage at such condition is known as the flat band voltage (V_{FB}). The other important parameter characterizing the MOS-capacitor is the inversion voltage (V_i). This occurs when the majority carriers in the semiconductor are inverted into the opposing type.

Figure 4.30 displays an example of the effects of γ-irradiation on the C-V_G characteristics. Both the V_{FB} and V_i experienced shifts toward the negative gate voltage when exposed to γ-rays. The values of $\Delta V_{FB} = 2V$ and $V_i = 2.25V$ were

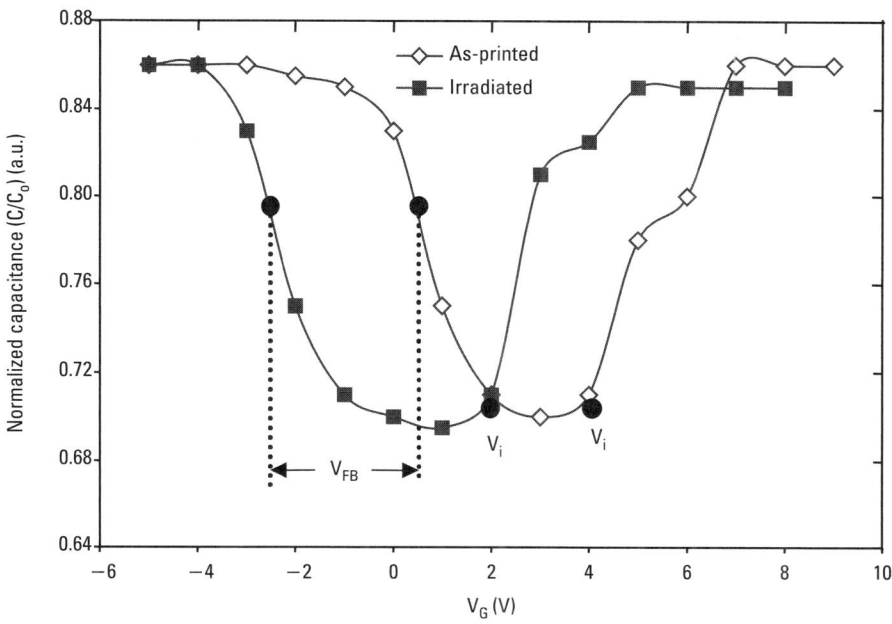

Figure 4.30 C-V characteristics of a MOS capacitor displays the shift toward a negative gate bias due to an exposure to a γ-ray dose of 180 Gy. (*From:* [65]. © 2004 Oxford University Press. Reprinted with permission.)

obtained when the Ag/Bi$_2$O$_3$/CuPc/Ag was exposed to a dose of 180 Gy [65]. Its C-V_{GS}, I_{DS}-V_{GS}, and I_{DS}-V_{DS} characteristics reflected the behavior of an enhancement-mode transistor. The threshold voltage was found to be 4.25V, which displayed a linear and gradual increase with increasing γ-ray dose.

In an analogous way, the I_{DS}-V_{GS} curves displayed in Figure 4.31 describe the effects that are encountered when the MOS transistor is exposed to γ-rays. The V_T is observed to decrease with an amount of $\Delta V_T \approx 2.25V$ when exposed to 180 Gy [65]. The shifts of the V_{FB}, V_i, and V_T toward the negative gate bias are attributed to the creation of radiation-induced traps (holes) in the bismuth oxide layer, which, in turn, induce negative image charges in both the gate (Ag) and the semiconductor (CuPc thick film). This results in an increase in the n-type conductivity, which causes a negative shift in the flat band voltage, the inversion voltage, and the threshold voltage.

Figure 4.32 shows two plots that describe the changes in the threshold voltage as a function of γ-ray dose for V_{DS} values of 0V and 2V [40]. The device was tested for a dose range up to 500 Gy. Two plots showed slopes of 8.3 mV/Gy for $V_{DS} = 0V$ and 16.8 mV/Gy for $V_{DS} = 2V$. The linearity exhibited by the threshold voltage as a function of γ-ray dose in this range indicated the

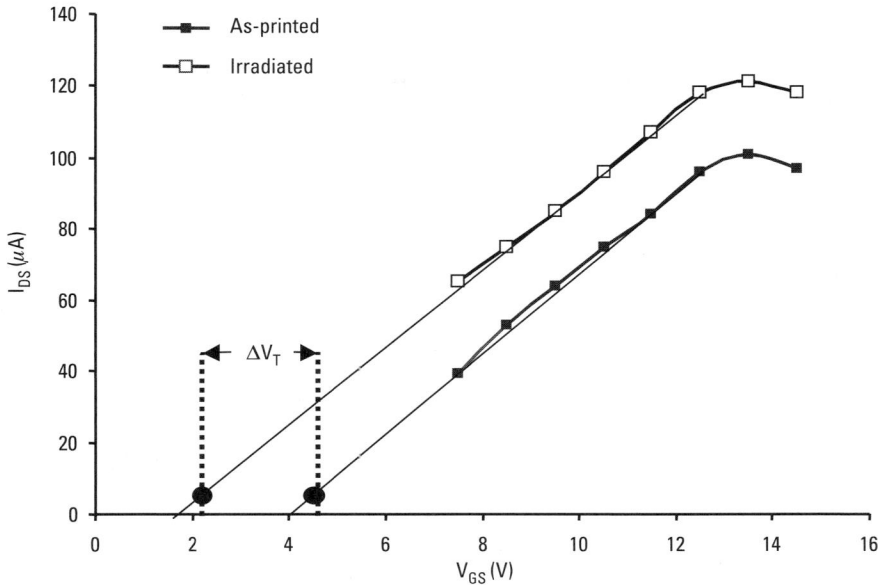

Figure 4.31 I_{DS}-V_G characteristics show a shift in the threshold voltage toward a negative gate bias caused by an exposure to a γ-ray dose of 180 Gy. (*From:* [65]. © 2004 Oxford University Press. Reprinted with permission.)

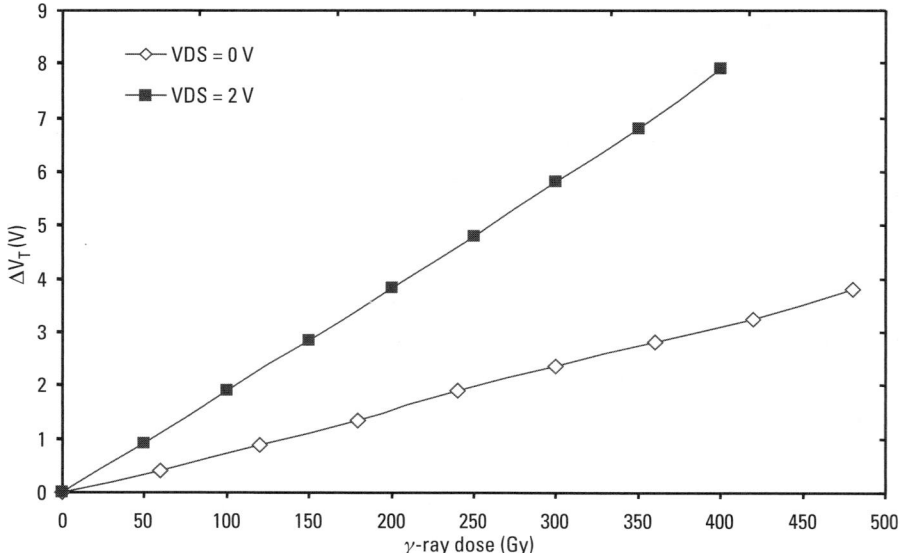

Figure 4.32 Threshold voltage versus γ-ray dose for $V_{DS} = 0V$ and $V_{DS} = 2V$. (*From:* [65]. © 2004 Oxford University Press. Reprinted with permission.)

suitability of the MOS device and its functional materials for γ-ray dosimetry applications.

4.4 Conclusion

The possibility of using films of metal oxides and metal-substituted phthalocyanines for radiation detection and dosimetry applications was widely explored. Both thin and thick film technologies were used for radiation sensor development. Doping of these pastes with carbon powder and mixing oxide materials in different proportions was found to control the sensitivity of these devices to radiation. Both thin and thick film devices were made in the form of resistor- and capacitor-type structures, planar structures with interdigitated electrodes, and p-n junctions. The ^{60}Co and ^{137}Cs sources were used to expose the samples to γ-radiation. A number of techniques were used for the characterization of electrical, optical, and structural properties of the devices. These include impedance spectroscopy; I-V characteristics; Hall effect measurements; UV-VIS spectroscopy; and SEM, XRD, and Raman spectroscopy.

In general, thin film devices were more sensitive to lower doses of radiation than their thick film counterparts. It is therefore recommended to use thin film devices for low-dose applications and thick film devices for higher dose applications, as the latter was found to sustain higher radiation doses. Detection of radiation was based on the fact that electrical, optical, and structural properties of the materials undergo changes under the influence of gamma radiation. The influence of radiation depends on both the dose and the parameters of the films, including their thickness: the degradation is more severe for the higher dose and the thinner films. Values of radiation damage in the samples were estimated from changes in their current-voltage characteristics and the optical absorption spectra, which were recorded after each exposure procedure. Annealing was found to restore both the electrical and the optical properties of thick film devices after they being damaged on γ-radiation exposure. This creates the ability to use them on repeatable bases. Thin film devices can hardly undergo annealing procedure, because their structures are too sensitive and diffusion of the electrode material may occur. Up to a certain level of radiation dose, all the samples showed an increase in the values of current. The regions of linear response may be considered a working region for dosimetry purposes, as usually radiation sensors have linear dose-response characteristics in certain region of doses. However, all the samples were susceptible to the environmental conditions, such as temperature, humidity, and electromagnetic field. This may result in under- or overestimation of the absorbed dose. To eliminate such effects, there is a need for corresponding signal conditioning.

The ability to fabricate a device that would satisfy the requirement of particular application, such as the sensitivity to γ-radiation exposure and working dose region, was demonstrated. The manufacture of thin/thick film structures should be considered a cost-effective alternative to the more traditional wafer-based fabrication techniques to produce sensors for real-time measurement of γ-radiation in dosimetry applications at room temperature. Acceptable resolution and performance of the proposed thin/thick film–based radiation sensors makes them promising alternatives for personal usage in emergency situation. It is believed that this novel approach will contribute toward cost-effective personnel dosimetry.

Since it is difficult to develop one sensor that can detect both low and high doses of ionizing radiation, corresponding devices can be fabricated for each range of interest. However, to cover more than one energy or type of radiation, the approach of using devices with a combined structure, such as sensor arrays, can be utilized, where sections of the radiation sensor could differ in material thickness or composition. These sensor arrays systems are discussed in detail in the next chapter.

References

[1] Bunshah, R. F., *Handbook of Deposition Technologies for Films and Coatings: Science, Technology, and Applications,* 2nd ed., Park Ridge, NJ: William Andrew/Noyes, 1994.

[2] Chapman, B. N., *Glow Discharge Processes: Sputtering and Plasma Etching,* New York: Wiley, 1980.

[3] Stuart, R. V., *Vacuum Technology, Thin Films, and Sputtering: An Introduction,* New York: Academic, 1983.

[4] Maissel, L. I., and R. Glang, *Handbook of Thin Film Technology,* New York: McGraw-Hill, 1983.

[5] Pan, J., G. L. Tonkay, and A. Quintero, "Screen Printing Process Design of Experiments for Fine Line Printing of Thick Film Ceramic Substrates," *Journal of Electronics Manufacturing,* Vol. 9, 1999, pp. 203–213.

[6] Jones, R. D., *Hybrid Circuit Design and Manufacture,* New York: Marcel Dekker, 1982.

[7] Harper, C. A., *Handbook of Thick Film Hybrid Microelectronics: A Practical Sourcebook for Designers, Fabricators, and Users,* New York: McGraw-Hill, 1997.

[8] Pacchioni, G., "Ab Initio Theory of Point Defects in Oxide Materials: Structure, Properties, Chemical Reactivity," *Solid State Sciences,* Vol. 2, No. 2, 2000, pp. 161–179.

[9] Heinrich, V. E., and P. A. Cox, *The Surface Science of Metal Oxides,* Cambridge, U.K.: Cambridge University Press, 1994.

[10] Tilley, R. J. D., *Principles and Applications of Chemical Defects,* Cheltenham, U.K.: Stanley Thornes, 1998.

[11] Colby, E., et al., "Gamma Radiation Studies on Optical Materials," *IEEE Trans. on Nuclear Science,* Vol. 49, No. 6, 2002, pp. 2857–2867.

[12] Phillips, G. W., et al., "Observation of Radiation Effects on Three-Dimensional Optical Random-Access-Memory Materials for Use in Radiation Dosimetry," *Applied Radiation and Isotopes,* Vol. 50, No. 5, 1999, pp. 875–881.

[13] Shpotyuk, O. I., "Amorphous Chalcogenide Semiconductors for Dosimetry of High-Energy Ionizing Radiation," *Radiation Physics and Chemistry,* Vol. 46, No. 46, 1995, pp. 1279–1282.

[14] Arshak, K. I., C. A. Hogarth, and M. Ilyas, "A Study of Electron Spin Resonance and Optical Absorption Edge in Amorphous Mixed Films of SiO and In_2O_3," *J. Mater. Sci. Lett.,* Vol. 3, 1984, pp. 1035–1038.

[15] Tominaga, K., et al., "Conductive Transparent Films Deposited by Simultaneous Sputtering of Zinc-Oxide and Indium-Oxide Targets," *Vacuum,* Vol. 59, No. 2–3, 2000, pp. 546–552.

[16] Arshak, K., and O. Korostynska, "Effect of Gamma Radiation onto the Properties of TeO_2 Thin Films," *Microelectronics International,* Vol. 19, No. 3, 2002, pp. 30–34.

[17] Arshak, K., and O. Korostynska, "Gamma Radiation-Induced Changes in the Electrical and Optical Properties of Tellurium Dioxide Thin Films," *IEEE Sensors,* Vol. 3, No. 6, 2003, pp. 717–721.

[18] Mott, N. F., and E. A. Davis, *Electronic Process in Non-Crystalline Materials,* Oxford, U.K.: Clarendon, 1979.

[19] Kurik, M. V., "Urbach Rule," *Physica Status Solidi,* Vol. A8, 1971, pp. 9–45.

[20] Tolpygo, S. K., et al., "Tc Enhancement by Low Energy Electron Irradiation and the Influence of Chain Disorder on Resistivity and Hall Coefficient in $YBa_2Cu_3O_7$ Thin Films," *Physica C: Superconductivity,* Vol. 269, No. 3–4, 1996, pp. 207–219.

[21] Croitoru, N., et al., "Influence of Damage Caused by Kr Ions and Neutrons on Electrical Properties of Silicon Detectors," *Nuclear Instruments and Methods in Physics Research Section A: Accelerators, Spectrometers, Detectors and Associated Equipment,* Vol. 426, No. 2–3, 1999, pp. 477–485.

[22] Hunter, L. P., *Handbook of Semiconductor Electronics,* New York: McGraw-Hill, 1970.

[23] Clough, R. L., "High-Energy Radiation and Polymers: A Review of Commercial Processes and Emerging Applications," *Nuclear Instruments and Methods in Physics Research Section B: Beam Interactions with Materials and Atoms,* Vol. 185, No. 1–4, 2001, pp. 8–33.

[24] Zaykin, Y., and B. A. Aliyev, "Radiation Effects in High-Disperse Metal Media and Their Application in Powder Metallurgy," *Radiation Physics and Chemistry,* Vol. 63, No. 3–6, 2002, pp. 227–230.

[25] Zhu, R. Y., "Radiation Damage in Scintillating Crystals," *Nuclear Instruments and Methods in Physics Research Section A: Accelerators, Spectrometers, Detectors and Associated Equipment,* Vol. 413, No. 2–3, 1998, pp. 297–311.

[26] Arshak, K., and O. Korostynska, "Influence of Gamma Radiation on the Electrical Properties of MnO and MnO/TeO_2 Thin Films," *Annalen der Physik,* Vol. 13, No. 1–2, 2004, pp. 87–89.

[27] Atanassova, E., et al., "Influence of γ-Radiation on Thin Ta_2O_5-Si Structures," *Microelectronics Journal,* Vol. 32, No. 7, 2001, pp. 553–562.

[28] Arshak, K., O. Korostynska, and J. Henry, "Structural Modifications in Thin Films Caused by Gamma Radiation," *Materials Science Forum,* Vol. 480–481, 2005, pp. 13–20.

[29] Champarnaud-Mesjard, J. C., et al., "Crystal Structure, Raman Spectrum and Lattice Dynamics of a New Metastable Form of Tellurium Dioxide: γ-TeO$_2$," *Journal of Physics and Chemistry of Solids*, Vol. 61, No. 9, 2000, pp. 1499–1507.

[30] Mirgorodsky, A. P., et al., "Dynamics and Structure of TeO$_2$ Polymorphs: Model Treatment of Paratellurite and Tellurite; Raman Scattering Evidence for New γ- and δ-Phases," *Journal of Physics and Chemistry of Solids*, Vol. 61, No. 4, 2000, pp. 501–509.

[31] Arshak, K., O. Korostynska, and J. Henry, "Thin Films of In$_2$O$_3$/SiO as Optical Gamma Radiation Sensors," *Proc. SPIE, Hard X-Ray and Gamma-Ray Detector Physics V*, 2004, pp. 83–91.

[32] Arshak, K., and O. Korostynska, "Thin Film Pn-Junctions Based on Oxide Materials as γ-Radiation Sensors," *Sensors and Actuators A: Physical*, Vol. 113, No. 3, 2004, pp. 307–311.

[33] Jonscher, A. K., and N. Siddiqui, "Decay of Photovoltage of Junction Diodes," *Solid-State Electronics*, Vol. 34, No. 11, 1991, pp. 1201–1206.

[34] Arshak, K., and O. Korostynska, "Radiation-Induced Changes in Thin Film Structures," *IEE Proc. Circuits, Devices and Systems*, Vol. 150, No. 4, 2003, pp. 361–366.

[35] Esaki, L., and Y. Miyahara, "A New Device Using the Tunneling Process in Narrow P-N Junctions," *Solid-State Electronics*, Vol. 1, No. 1, 1960, pp. 13–14.

[36] Engel, M., "Single-Crystal and Solid-State Molecular Structures of Phthalocyanine Complexes," *Technical Report*, 1996, pp. 11–54.

[37] Ambily, S., and C. S. Menon, "Determination of the Thermal Activation Energy and Optical Band Gap of Cobalt Phthalocyanine Thin Films," *Materials Letters*, Vol. 34, No. 3–6, 1998, pp. 124–127.

[38] Gould, R. D., "Structure and Electrical Conduction Properties of Phthalocyanine Thin Films," *Coordination Chemistry Reviews*, Vol. 156, 1996, pp. 237–274.

[39] Arshak, A., et al., "Effect of γ-Radiation on the Conduction Mechanism of Thermal Vacuum Deposited Copper Phthalocyanine Thin Films," *Proc. IEEE-NANO*, 2001, pp. 238–242.

[40] Arshak, K., et al., "Thin and Thick Films of Metal Oxides and Metal Phthalocyanines as Gamma Radiation Dosimeters," *IEEE Trans. on Nuclear Science*, Vol. 51, No. 5, 2004, pp. 2250–2255.

[41] Gould, R. D., and A. K. Hassan, "High Field Electronic Conduction Processes in Copper Phthalocyanine Thin Films Using Lead and Gold Electrode Combinations," *Thin Solid Films*, Vol. 194, No. 1–2, Pt. 2, 1990, pp. 895–904.

[42] Abdel-Malik, T. G., "Schottky Barrier Formation and Transport Properties in Copper Phthalocyanine Dispersed in Polycarbonate," *Proc. International Society for Optical Engineering*, Denver, CO, 2004, pp. 136–141.

[43] Arshak, K., and R. Perrem, "The Correlation of the Electrical and Optical Properties of Thin Films of V$_2$O$_5$-Bi$_2$O$_3$," *Journal of Physics D: Applied Physics*, Issue 7, 1993, pp. 1098–1102.

[44] Jones, B. E., and T. O. Marshall, "The Dosimetry of Mixed Radiations, Involving More Than One Energy or Type of Radiation, with the R.P.S./A.E.R.E. Film Dosimeter," *The Journal of Photographic Science*, Vol. 12, 1964, pp. 319–327.

[45] Lee, S. Y., and K. J. Lee, "Development of a Personal Dosimetry System Based on Optically Stimulated Luminescence of α-Al$_2$O$_3$:C for Mixed Radiation Fields," *Applied Radiation and Isotopes*, Vol. 54, No. 4, 2001, pp. 675–685.

[46] Arshak, K., O. Korostynska, and J. Harris, "γ-Radiation Dosimetry Using Screen Printed Nickel Oxide Thick Films," *Proc. 23rd Intl. Conf. on Microelectronics (MIEL)*, Nis, Serbia, May 12–15, 2002, pp. 357–360.

[47] Pejova, B., et al., "A Solution Growth Route to Nanocrystalline Nickel Oxide Thin Films," *Applied Surface Science*, Vol. 165, No. 4, 2000, pp. 271–278.

[48] Arshak, K., O. Korostynska, and F. Fahim, "Various Structures Based on Nickel Oxide Thick Films as Gamma Radiation Sensors," *Sensors*, Vol. 3, 2003, pp. 176–186.

[49] Balberg, I., "A Comprehensive Picture of the Electrical Phenomena in Carbon BlackPolymer Composites," *Carbon*, Vol. 40, 2002, pp. 139–143.

[50] Grimmett, G., *Percolation*, Berlin: Springer-Verlag, 1999.

[51] Vilcáková, J., P. Sáha, and O. Quadrat, "Electrical Conductivity of Carbon Fibres/Polyester Resin Composites in the Percolation Threshold Region," *European Polymer Journal*, Vol. 38, No. 12, 2002, pp. 2343–2347.

[52] Celzard, A., et al., "Electrical Conductivity of Carbonaceous Powders," *Carbon*, Vol. 40, 2002, pp. 2801–2815.

[53] McLachlan, D. S., K. F. Cai, and G. Sauti, "AC and DC Conductivity-Based Microstructural Characterization," *International Journal of Refractory Metals and Hard Materials*, Vol. 19, 2001, pp. 437–445.

[54] McLachlan, D. S., M. Blaszkiewicz, and R. E. Newnham, "Electrical Resistivity of Composites," *Journal of American Ceramic Society*, Vol. 73, 1990, pp. 2187–2203.

[55] Korostynska, O., K. Arshak, and M. Mahon, "Gamma Radiation Sensing Properties of NiO Thick Film Pn-Junctions," *Proc. IEEE Sensors*, 2003, pp. 79–83.

[56] Kamiya, M., et al., "Intrinsic and Extrinsic Oxygen Diffusion and Surface Exchange Reaction in Cerium Oxide," *Journal of The Electrochemical Society*, Vol. 147, 2000, pp. 1222–1227.

[57] Garvie, L. A. J., and P. R. Buseck, "Determination of Ce^{4+}/Ce^{3+} in Electron-Beam-Damaged CeO_2 by Electron Energy-Loss Spectroscopy," *Journal of Physics and Chemistry of Solids*, Vol. 60, No. 12, 1999, pp. 1943–1947.

[58] Arshak, K., and O. Korostynska, "γ-Radiation Sensing Properties of Cerium Oxide Based Thick Film Structures," *Sensors and Actuators A: Physical*, Vol. 115, No. 2–3, 2004, pp. 196–201.

[59] Arshak, K., O. Korostynska, and J. Henry, "Thick Film Pn-Junctions Based on Mixed Oxides of Indium and Silicon as Gamma Radiation Sensors," *Microelectronics International*, Vol. 21, No. 1, 2004, pp. 19–27.

[60] Arshak, A., et al., "Effect of γ-Rays on the Optical and Electrical Properties of Copper-Phthalocyanine Thick Films," *Proc. 23rd Intl. Conf. on Microelectronics (MIEL)*, Nis, Serbia, May 12–15, 2002, pp. 353–356.

[61] Arshak, A., et al., "Investigations of the Effects of Gamma-Radiations on the Optical and Electrical Properties of Nickel Phthalocyanine (NiPc) Thick Film, International Microelectronics and Packaging Society," *Proc. 35th International Symposium on Microelectronics Proceedings*, Denver, CO, September 4–6, 2002, pp. 760–765.

[62] Arshak, A., S. M. Zleetni, and K. Arshak, "γ-Radiation Sensor Using Optical and Electrical Properties of Manganese Phthalocyanine (MnPc) Thick Film," *Sensors*, Vol. 2, 2002, pp. 174–184.

[63] Arshak, A., et al., "High Dose Optical and Electrical Sensor Dosimeter Using Cobalt Phthalocyanine (CoPc) Thick Film, International Microelectronics and Packaging Society," *Proc. 35th International Symposium on Microelectronics Proceedings,* Denver, CO, September 4–6, 2002, pp. 766–771.

[64] Holmes-Siedle, A. G., and L. Adams, *Handbook of Radiation Effects,* New York: Oxford University Press, 1993.

[65] Arshak, A., S. Zleetni, and K. I. Arshak, "MOS Dosemeter Using Bismuth Oxide (Bi_2O_3) and Copper Phthalocyanine (CuPc) Polymer Thick Film," *Radiation Protection Dosimetry,* Vol. 111, No. 1, 2004, pp. 77–81.

[66] Soliman, F. A. S., et al., "Characteristics and Radiation Effects of MOS Capacitors with Al_2O_3 Layers in P-Type Silicon," *Applied Radiation and Isotopes,* Vol. 46, No. 5, 1995, pp. 355–361.

5

Sensor Arrays, Radiation Nose Concept, and Pattern Recognition[1]

5.1 Sensor Arrays

5.1.1 In_2O_3/SiO Sensor Array

As was shown in Chapter 4, the sensitivity of metal oxide and polymer films to γ-radiation exposure depends on their composition and thickness. Mixing materials in different proportions and the addition of conducting particles, such as carbon, alters the film's susceptibility to radiation. The level of sensitivity and working dose range of such sensors is conditioned by material properties and the device structure, including the thickness of the radiation-sensitive layer. Therefore, a combination of a number of sensors with different response parameters into sensor arrays would enhance the overall performance of the radiation detection system.

To illustrate the effect of composition, Figure 5.1 shows the dependences of normalized values of currents versus radiation dose for four thick film radiation sensors with different In_2O_3/SiO constituents, which are built in to a sensor array [1]. They were discussed in detail in Section 4.3.1.4.

Figure 5.1 illustrates that all samples showed a large increase in the values of current up to a dose of 114 μGy, except pure In_2O_3 samples (170 μGy). Beyond these doses, the values of normalized current were found to be highly

---

1. In cooperation with C. Cunniffe.

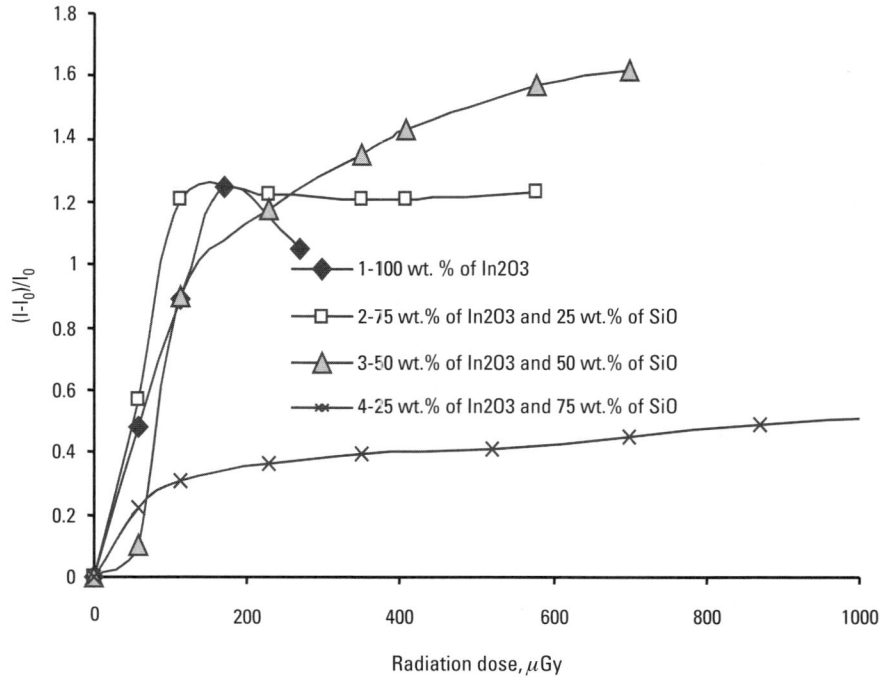

Figure 5.1 Dependences of normalized values of currents versus radiation dose for four sensors with different In_2O_3/SiO constituents. (*From:* [1]. © 2006 IEEE. Reprinted with permission.)

dependent on the material composition. The threshold levels of radiation after which samples were damaged were:

- 170 μGy for films made with 100 wt. percent of In_2O_3;
- 578 μGy for films made with 75 wt. percent of In_2O_3 and 25 wt. percent of SiO;
- 700 μGy for the films made with 50 wt. percent of In_2O_3 and 50 wt. percent of SiO;
- 2,100 μGy for the films made with 25 wt. percent of In_2O_3 and 75 wt. percent of SiO.

The most important aspect of utilizing multiple radiation sensors is choosing the most accurate one for a given dose range. Additionally, the detection of damaged sensors is a critical task necessary for ensuring the maximum possible accuracy in measuring radiation dose. The characteristic of a radiation sensor exposed to a radiation dose higher than its working dose range will permanently change, making the sensor unreliable. Therefore, damaged sensors should be excluded from further usage. For example, a sensor based on pure In_2O_3 is damaged if exposed to a radiation dose level higher than 170 μGy.

The process of the detection always starts from analysis of radiation dose readings from sensor 4, which is capable of measuring the highest radiation doses without being damaged [1]. If sensor 4 reports the detection of radiation dose of more than 700 μGy, then sensors 1, 2, and 3 should be classified as damaged. When sensors 3 and 4 report radiation dose higher than 578 μGy, then sensors 1 and 2 are damaged. Analysis starts from sensor 4, and if it does not detect damage in sensor 3, then sensor 3 is used (as it is more accurate) to detect whether sensors 1 or 2 are damaged. Based on this example, one can clearly see the effect of composition on the sensitivity of material to radiation.

5.1.2 PolyVinyliDene Fluoride Films of Different Thicknesses

PVDF is a long chain high-molecular-weight polymer with predominant repeating unit established as $(-CH_2-CF_2-)$. This semicrystalline material with a melting point of 338°C is widely used for various technical/engineering applications. Several types of sensors based on PVDF films were designed and tested, including laser radiation sensors, a light pressure transducer, an ion radiation sensor, and a multifunctional alarm [2]. Semicrystalline PVDF, irradiated with different ions in the range 100 keV–100 MeV, exhibited a strong change in the crystalline structure, as detected by X-ray diffraction measurements and calorimetric analysis [3]. At ion doses higher than 300 kGy, PVDF showed a degradation in the crystalline structure, reduction of hydrogen and fluorine concentration, and the formation of carbon reticulation. The relationship between the crystallinity of PVDF and electron radiation effects was studied with differential scanning calorimetry and X-ray diffraction [4]. It was reported that at doses lower than 400 kGy, the crystallinity of PVDF increased slightly with the increase in absorbed dose.

To examine the effect of material thickness on the sensitivity of the sensors to γ-radiation, PVDF thick-film structures were manufactured and tested. To increase the conductivity of the films, they were filled with 6 wt. percent of carbon. At a fixed applied voltage of 3V, a tenfold increase in the value of current was recorded after a dose of 228 μGy for PVDF films with a thickness of 23.97 μm. Thicker films (39.12 μm) showed an increase in the values of current with irradiation to a dose of 798 μGy. To trace and compare the response to radiation of these two samples with different thicknesses, see Figure 5.2, which shows the dependence of normalized currents $(I-I_0)/I_0$ versus γ-dose. The conclusion can be drawn that the thicker films can sustain higher radiation doses, whereas thinner films are more sensitive to the lower doses, although they are damaged sooner, which is in agreement with the results reported earlier [5, 6].

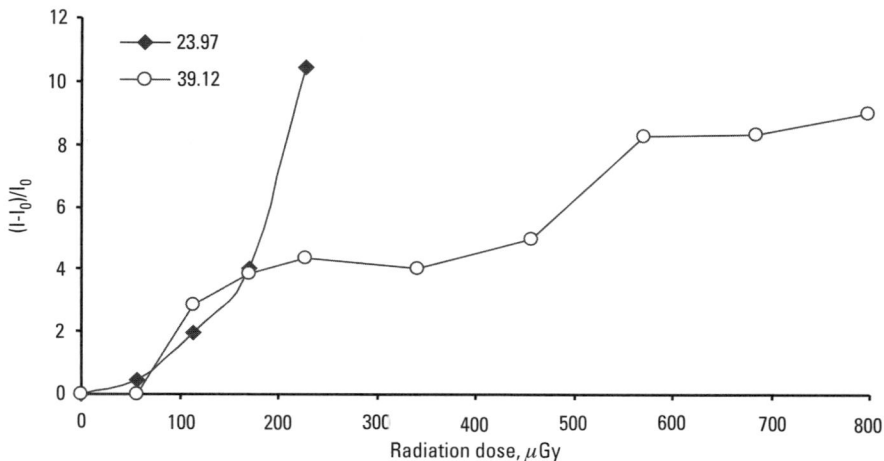

Figure 5.2 Dependences of normalized values of currents versus γ-dose for PVDF sensors with different thicknesses.

5.1.3 Multiple Sensor Materials Incorporated into an Array

Although mixing materials in different proportions and manipulating the thickness of the sensing layer can improve the sensitivity and to a certain degree extend the working dose range of dosimeters, different materials must be used when the targeted range of measurements is very broad. Cobalt phthalocyanine (CoPc) is one of the materials suitable for high-dose dosimetry. Screenprinted CoPc films with a thickness of 15 μm showed an increase in the values of current with the increase in gamma dose to 27 kGy [7]. For example, one may combine PVDF (μGy) and CoPc (kGy) sensors into single array to achieve superior sensitivity to radiation over a broad dose range.

5.2 Dosimetry of Mixed Radiations

The previously proposed radiation sensor arrays have two main advantages (i.e., wider working dose range and increased overall sensitivity). In view of the wide variety of radiation sources employed in industry, medicine, and research, it is desirable that a device used for dosimetry be capable of giving a reasonably accurate estimate of the total dose received from exposure to radiations differing in both energy and type. It is also necessary to be able to separate the individual contributions of the different types of radiation that make up the mixture, since the permissible levels for α-, β-, and γ-radiation differ appreciably. Such knowledge is particularly essential if any attempt is made to determine the dose delivered at a depth within a body. It may be of value in cases where an investigation is necessary to determine the cause of an overexposure.

Potential applications of the dosimetry of mixed radiations include cases when type and level of radiation is unknown, such as building sites, mining, military, and security purposes. When developing a new construction site, builders could face granite or marble with unknown radiation levels, not to mention radon gas, the natural source of α-radiation. Miners could face similar situation at their work as well, where it will be useful to identify the type and levels of radiations that might be present. When the presence of complex mixtures of radiations of various types and energies is expected, one can use a range of dosimeters, each designed for a certain energy range. However, in an unexpected situation or in cases of nuclear emergency, which were discussed in Section 2.7, there is a need for a compact real-time dosimeter of mixed radiation field, with visible/audio alarm signal if the safe dose is exceeded.

When measured radiation exposure is of a complex nature (e.g., includes α-, β-, and γ-rays), the radiation-sensitive materials should be carefully chosen. Similarly, an array of sensors can be implemented, with each sensor sensitive (ideally) to only one type of radiation. This can be achieved via corresponding shielding, where each sensor in the array is covered with a certain material that allows only one type of radiation to be picked up by that sensor.

5.2.1 Electronic Nose Systems

This concept is akin to the human olfactory system, where the odor passing through the nasal passage interacts with the olfactory receptacles, which transmit a signal to the olfactory bulb and onto the brain [8], hence the name radiation nose. The objective of an electronic nose, sometimes called e-nose, system is to mimic the human olfactory structure to enable artificial detection of odors and odor concentration. The term *electronic nose* was initially used to describe this instrument by Gardner in the late 1980s [9]. Before this time, these sensors were referred to as gas sensors and were first developed in 1954 by Hartman [10, 11]. The electronic nose finds its applications not only in the food industry, but also in healthcare, environmental monitoring, agriculture, military, and space exploration. Within the food industry, the electronic nose has been utilized to determine freshness of foods stuffs such as fish [12], discern the ripeness of fruits [13], distinguish between coffee beans [14], and distinguish between alcohols [15]. In the health care industry, there is an ongoing research into noninvasive techniques to detect illness in the human. Using the electronic nose to sample the subjects' breath and detect volatile markers in the breath can indicate the presence of illnesses [16] such as lung cancer [17, 18], breast cancer [19], and respiratory conditions [20]. Within the manufacturing industry, the quality of commercial cardboard paper was examined with electronic nose system [21]. By analyzing the vapors over the milk of dairy cows, early detection of a condition called ketosis can be detected [22]. Environmental monitoring tasks, such as detecting pollutants in domestic wastewater [23] and monitoring the quality of potable water [24], are also useful applications of e-nose systems.

One important application for the electronic nose is for use in detection of explosives. An electronic nose capable of detecting explosive may be used for the detection of landmines and for homeland security purposes. Homeland security applications include screening people's packages, luggage, and vehicles at key locations such as airports or government buildings for the prevention of terrorist attacks [25]. There are about 120 million unexploded landmines remaining across the world in 70 countries, killing or maiming 26,000 people per year [8, 26]. At present, the detection of landmines is carried out using either dogs for sniffing or metal detectors. Both of these techniques have their drawbacks, which would be addressed by the use of an electronic nose. The techniques require a human to either command the dog or control the metal detector, putting the human at risk. Also with the metal detector, technique can be slow and dangerous—primarily because some landmines contain little or no metal, and secondly because there can be a high false alarm rate due to scattered metal. Using dogs to detect landmines can also be slow, as they have a short attention span and tend to lose their concentration after 30–120 minutes [8, 26]. Near landmines, either low concentrations of gasses or particles of explosives can be detected. It has been observed that with 2,4,6-trinitrotoluene (TNT), the order of a few nanogramms of explosive particles are present in the vicinity of a landmine containing TNT. Materials that are of interest to detect are TNT and 1,3,5-trinitro-1,3,5-triazocyclohexane (RDX), which are in the most commonly used landmines [8]. Using an electronic nose to detect landmines and explosives would not require human control in the vicinity of detector. The device could be either autonomous or remotely controlled by an operator.

A variety of sensors are currently used in e-noses. These sensors work on the physical properties of conductivity, piezoelectricity, capacitive-charge coupling, fluorescence, chemo-luminescence, molecular spectrum, atomic mass spectrum, and transmitted light spectrum [14]. Table 5.1 outlines names and Web pages of companies that manufacture and sell commercial e-nose systems.

An analysis of gas mixtures requires the use of multiple gas sensors, each one responding by varying degrees to different gases or gas mixtures. This analogy is applicable to the area of radiation sensors. As an alternative to using many different dosimeters to detect mixed radiations of different dose rates/types, an array of sensors may be constructed. The resulting sensors array response yields a pattern based on the varying responses of the individual sensor in the array. The main advantage of such a system is that the sensors are integrated onto one device, which is capable of distinguishing between different sources of radiation and varying dose rates.

Table 5.1
Commercially Available Electronic Nose Systems

Agilent Technology	http://www.agilent.com/
Airsense	http://www.airsense.com/uk/main.htm
Alpha MOS	http://www.alpha-mos.com/
AppliedSensor	http://www.appliedsensor.com/
Scensive Technologies	http://www.scensive.com/
Chemsensing, Inc.	http://www.chemsensing.com/
Cyrano Sciences	http://www.cyranosciences.com/
Environics Industry Oy	http://www.environics.fi/
Estcal	http://www.estcal.com/
HKR Sensorsysteme	http://www.hkr-sensor.de/
Illumina, Inc.	http://www.illumina.com/
Lennartz Electronic	http://www.lennartz-electronic.de/
Microsensor Systems	http://www.microsensorsystems.com/
OligoSense	http://www.oligosense.be/
Osmetech plc	http://www.osmetech.plc.uk/
RST Rostock	http://www.rst-rostock.de/
Sensobi Sensoren	http://www.sensobi.com/
SMart Nose	http://www.smartnose.com/
Technobiochip	http://www.technobiochip.com/

5.3 Radiation Nose Concept: Overview of Pattern Recognition

5.3.1 Radiation Nose Concept

Figure 5.3 presents a concept image of a radiation nose system, where a single substrate is used for multiple sensors, with each sensor engineered to respond to one type of radiation (e.g., alpha, beta, gamma, proton or neutron).

When there is no radiation, an associated value of current is passing through each resistor-type sensor (e.g., I_1, I_2, I_3, and so forth, as shown in Figure 5.3). Once a mixed radiation field interacts with the sensor array, it induces physical and/or chemical changes in the sensing materials, which causes an associated change in the electrical properties (e.g., conductivity). Figure 5.4 demonstrates the response of a radiation nose system to a mixed radiation field comprising alpha, beta, and gamma radiations. Figure 5.5 shows the response of the same system, but in this case the mixed radiation field consists of gamma,

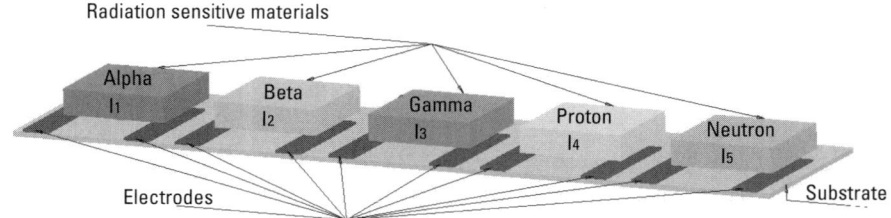

Figure 5.3 The concept of radiation nose system.

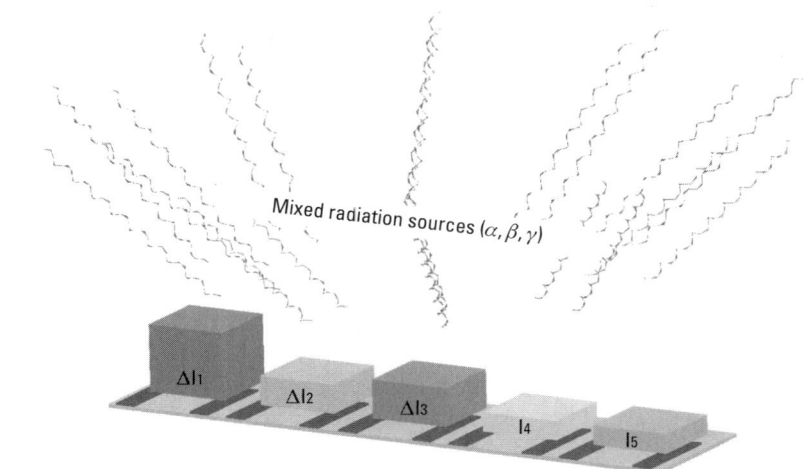

Figure 5.4 The response of a radiation nose system to a mixture of α-, β-, and γ-radiations.

proton, and neutron radiations. One may monitor a change in the values of current ΔI of associated sensor to determine the type and level of radiation. For example, when α-, β-, and γ-radiations interact with a radiation nose system (Figure 5.4), there are changes in the values of currents ΔI_1, ΔI_2, and ΔI_3, respectively, whereas no changes are recorded for sensors 4 and 5. When a mixture of gamma, proton, and neutron radiations is present, one may expect the response of three corresponding sensors, as shown in Figure 5.5.

As an example, Figures 5.4 and 5.5 showed the sensors array designed for five different radiation sources, with a single sensor for each type of source. However, multiple sensors for the same type of radiation may be accommodated to cover different ranges of radiation, as was shown in Section 5.1. Figure 5.6 illustrates general block diagram of a sensor arrays system.

In practice, when perfect shielding is difficult to implement, the same sensor could respond to a number of radiations, so cross-sensitivity occurs. For example, the decay of ^{117}Cs (gamma radiation) is often accompanied by around

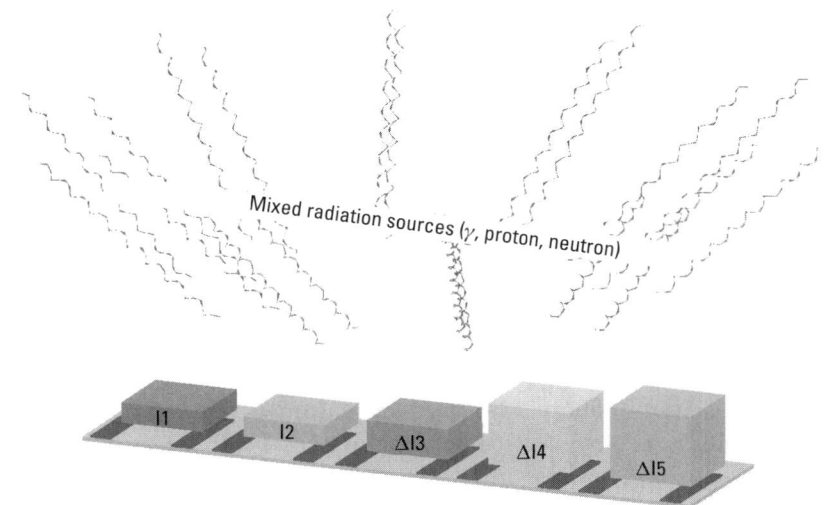

Figure 5.5 The response of a radiation nose system to a mixture of gamma, proton, and neutron radiations.

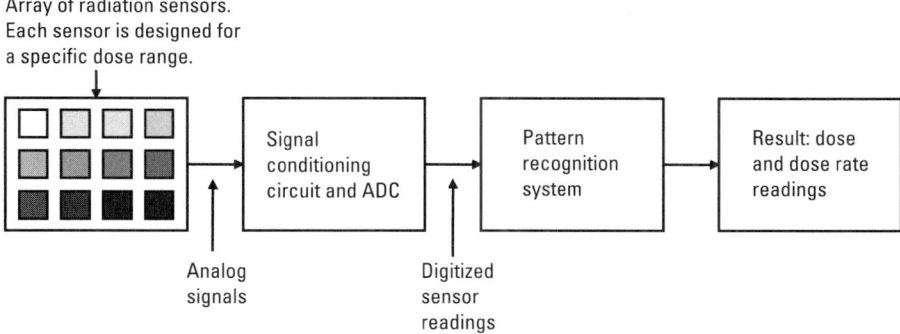

Figure 5.6 Block diagram of gamma radiation sensor arrays system.

8% of β-radiation. The sensors will respond to some extent to a radiation exposure, but a larger change could be exhibited in some sensors. It is these changes across multiple sensors that define the *fingerprint* pattern for a given type and energy of the source. The unique fingerprint is the raw data used for the PR algorithm.

Once the radiation has interacted with the sensor array, it is the function of the interface circuit to measure the changes in the physical properties of the sensor. The sensor array is interrogated by a signal conditioning circuit, which depends on the nature of the sensor (e.g., resistive, capacitive, diode). Subject to the electrical changes in the sensors being measured, the signal conditioning circuit may then be used to buffer, amplify, and filter the signal and execute some compensation tasks, such as linearization or temperature compensation [27]. The conditioned signal is digitized using an analog to digital converter (ADC) and may be analyzed using techniques described in the following sections.

5.3.2 Overview of Pattern Recognition

Based on a set of the output readings from the sensors, a specific PR algorithm should be applied, so that the radiation nose system will accurately determine radiation dose and type.

These radiation-induced changes in the sensors are transduced into electrical signals, which are preprocessed and conditioned before identification by a PR system. The e-nose systems are designed so that the overall response pattern from the array is unique for a given odor in a family of odors to be considered by the system [28]. The main goal of the PR algorithm is to classify a set of input values into one of the available classes. In the case of e-nose system, input values consist of readings from each sensor in the array, while a class recognized by the PR algorithm reflects a type of odor (type and dose in radiation nose systems). PR covers a very rich family of different algorithms. Performance of a particular algorithm can be measured in terms of flexibility, efficiency, and reliability. Training procedures for different PR algorithms may vary significantly, however the common goal is to create a classifier that correctly classifies all samples from a training data set. Accuracy of measurement classification by PR algorithms is the most common problem in real applications of e-nose.

It is anticipated that sensor arrays can determine not only the total received dose, but also can record a history of any changes in dose rate with time. This can help to determine the cause of radiation—whether it was a major short release of radiation or there is a leak in shielding somewhere, which has to be immediately localized. Radiation response is rate-dependent [29]; therefore, each sensor will react differently on sequence of doses received with time. Thus, the system consisting of a series of sensor arrays is capable of providing more detailed information than its single-sensor detector counterpart. Measurement of radiation, especially from few different sources or types, is typically affected by many factors: sensor nonlinearity, sensor drift, cross-sensitivity to multiple types/doses of radiation, accuracy, working dose range, and so forth. A dynamic selection of multiple sensors of various working dose ranges coupled with a carefully chosen PR analysis algorithm can maximize the accuracy of classification.

A PR system is comprised of many data manipulation blocks, as shown in Figure 5.7. Initially the raw data must be acquired, resulting in a time-dependent response curve when exposed to a source. Baseline manipulation is then executed

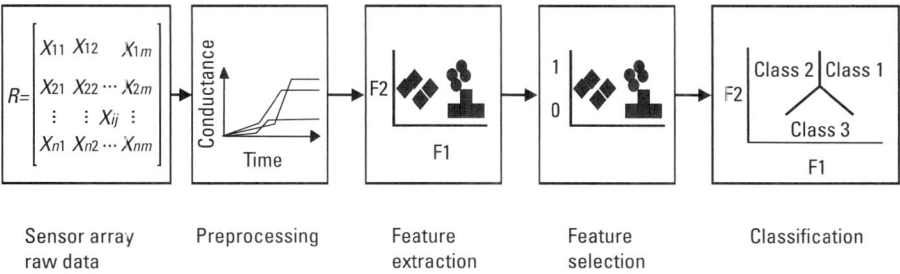

Figure 5.7 Block diagram of PR system.

on the raw data. Following this, feature extraction and then feature selection are implemented. Classification may then be carried out. Data analysis techniques may be categorized as quantitative-qualitative; model based-model free [30]; parametric-nonparametric; or supervised-unsupervised [8].

5.3.2.1 Preprocessing

Each sensor in the array will produce a time-dependent response $R_{ij}(t)$, where R is the response of sensor i to a source j at time t [8]. The number of sensors in the array defines the dimensionality of the sensor space [31]. Preprocessing is used to extract the required information from the response curves. Depending on the underlying technology of the sensor, the features that may be extracted are the maximum response, the rising slope of the curve, or the area under the curve.

5.3.2.2 Baseline Manipulation

Baseline manipulation transforms the sensor response relative to the baseline. The methods of baseline manipulation implemented in e-noses are differential, fractional, relative, and logarithm. These techniques may be used for contrast enhancement and drift compensation [32]. The following outlines ways to implement manipulation techniques, resulting in a time-independent response if the delta value is extracted or a time-dependent response if the techniques are applied to the full response curve.

In (5.1)–(5.4) R_{ij} is the acquired data from the sensor, which may be a conductance, resistance, voltage, current, or capacitance reading. R_{ij}^{min} is the minimum reading or the background reading before the sensor is exposed to a source, and X_{ij} is the new manipulated value.

The differential baseline manipulation shown in (5.1) subtracts the baseline and can help reduce the effects of additive drift [32]:

$$X_{ij} = R_{ij} - R_{ij}^{min} \qquad (5.1)$$

Relative baseline manipulation shown in (5.2) divides by the baseline, can help reduce the effect of multiplicative drift, and generates a dimensionless response [32]:

$$X_{ij} = R_{ij} / R_{ij}^{\min} \qquad (5.2)$$

Fractional baseline manipulation shown in (5.3) is a combination of the previous two methods, where the baseline is subtracted and then divided by the baseline resulting in a dimensionless normalized result [32]. This method was mainly used for presentation of data in normalized current versus radiation dose form (see Chapter 4):

$$X_{ij} = \frac{R_{ij} - R_{ij}^{\min}}{R_{ij}^{\min}} \qquad (5.3)$$

Logarithm shown in (5.4) is a less frequently implemented method and is used to linearize the sensor response for devices that can be modeled using the power law [31]:

$$X_{ij} = \log_{10}\left(\frac{R_{ij}}{R_{ij}^{\min}}\right) \qquad (5.4)$$

5.3.2.3 Standardization

Signals from the sensors with small variance are often as important as those from the sensors showing large variance. The variances of all sensors are made numerically equal using standardization [30]:

$$X'_{ij} = \frac{x_{ij} - \bar{x}}{\sigma} \qquad (5.5)$$

where $\bar{x}$ is the mean and σ is the standard deviation. $x_{ij} - \bar{x}$ centers the data, and dividing by σ standardizes the data.

5.3.2.4 Normalization

Normalization maps the response vectors onto a unit radius hypershere, resulting in a response vector of unit length, as expressed by [31]:

$$x'_{ij} = \frac{x_{ij}}{\sqrt{\sum_{k=1}^{m} x_{kj}^2}} \qquad (5.6)$$

where m is the number of sensors and j represents the source.

5.3.2.5 Scaling

Scaling is applied to the data to prepare it for use in neural networks, where the data must be scaled between the values -1 and $+1$, or for use in a Fuzzy ARTMAP system [33], where the data must be scaled between 0 and $+1$ [30].

5.3.3 Feature Extraction/Selection

Given a set of sensors sampled over time, the resulting data is potentially large. Feature extracting is concerned with extracting the most descriptive feature of that response, retaining the maximum information. The result is a single value for each sensor. The resulting feature set may also contain sensors that reached saturation point or do not contain valid information. Feature selection is a technique to remove the sensor values with no valid contribution, resulting in a feature subset of the original set [32].

5.3.3.1 Feature Extraction

Extracted features depend on the time-dependent response curve acquired from the sensor. Features extracted from response curve are the maximum change in response, the rise time, the area under the curve, or the rising slope of the curve. The result of using one of the feature-extraction techniques is a time-independent response vector in response to a radiation source j. Therefore a matrix of all time-independent vectors in response can be created (5.7).

$$R = \begin{bmatrix} X_{11} & X_{12} & \cdots & X_{1m} \\ X_{21} & X_{22} & \cdots & X_{2m} \\ \vdots & \vdots & X_{ij} & \vdots \\ X_{n1} & X_{n2} & \cdots & X_{nm} \end{bmatrix} \qquad (5.7)$$

where n is the number of time-independent response vectors in the matrix and m is the number of sensors.

Using the previous example with the array of four different In_2O_3/SiO sensors, the best feature to extract is the slope of the curves (Figure 5.1). In order to determine the best subset of sensors to select, it is necessary to check the value of the slope. As long as the slope is positive or greater that zero the feature is valid. If the slope is negative, the sensor has broken down and the feature may be removed.

5.3.3.2 Feature Selection

Feature selection is concerned with reducing the number of input variables to the classification algorithm. The benefits of this are a reduction in cost and time of processing the data, and irrelevant information can be eliminated [34]. Feature selection requires a search strategy and an objective function. The search strategy is used to search for the best possible combination of features to make up the final feature vector to be presented to the classifier [30]. Search methods such a sequential forward selection (SFS), sequential backward selection (SBF) and genetic algorithms (GA) may be used for further feature selection [35]. The objective function evaluates the extracted feature and returns a measure of the relevance of the feature; this result is used in selecting further features until the feature subset is selected. There are two types of objective functions: filters and wrappers. Filters compare feature subsets by their information content [32] by using distance metrics to measure class separation or correlation measures [30]. Wrappers evaluate the feature subsets based on their predictive accuracy in the classification [32].

5.4 Classification and Validation

5.4.1 Classification

Classification is a process where an unknown feature vector is presented to the classifier and assigned a class from a set of previously learned classes. In supervised classification techniques, a set of unknown feature vectors and the class to which they belong are presented to the classification algorithm. Following training, an unknown feature vector is presented to the classifier to be assigned a class. Statistical chemometrics are techniques based on principles of engineering, mathematics, and statistics, while biologically inspired methods are nonparametric techniques and to a certain extent mimic the human olfactory system. Figure 5.8 gives examples of some classification techniques used in e-nose applications and categorizes them based on their properties of operation [8]. The meanings of abbreviations and references for the detailed description of these methods are given in Table 5.2.

5.4.2 Principal Component Analysis

Although principal component analysis (PCA) is not strictly a PR algorithm, it is by far the most popular method used in nearly all electronic nose systems for data preprocessing and analysis. PCA is typically used to reduce data dimensionality

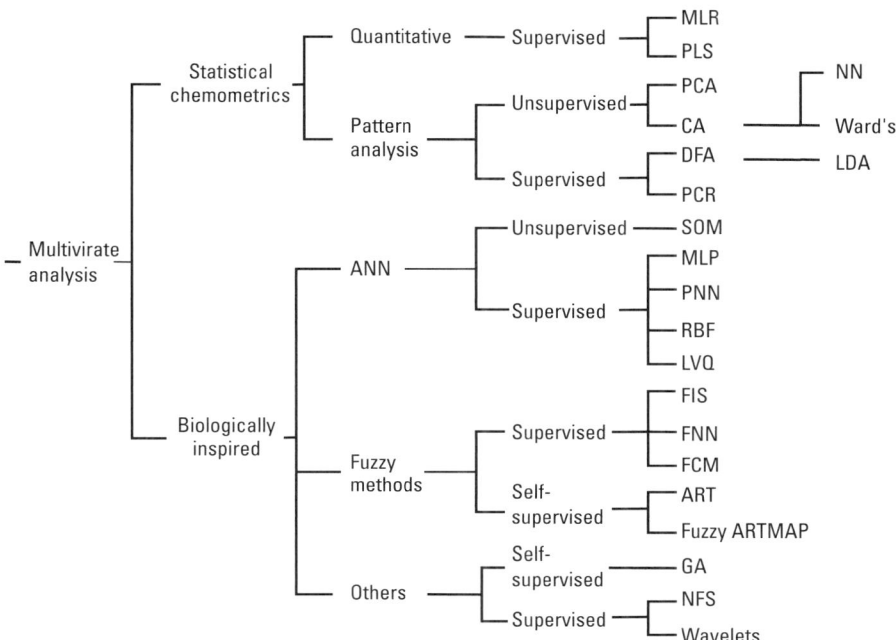

Figure 5.8 Classification scheme of the multivariate pattern analysis techniques. (*From:* [8]. © 2003 Wiley-VCH. Reprinted with permission.)

by eliminating redundant and correlated data from original set of samples without significant loss of the information [36]. Dimensionality reduction is especially important for data visualization. Most nose-type systems consist of more than three sensors, which normally would be difficult to plot on two- or three-dimensional graphs. PCA allows the conversion of highly dimensional data for visualization on classic two-dimensional plots. An example of such an application is presented in [20], which describes an evaluation of the application of e-nose systems to diagnose illness. Detection of the bacteria was performed by a 6-MOS gas sensor array and the PCA, which transformed sensor samples to two dimensions. The resulting data, visualized on a two-dimensional plot, indicates the presence of two distinct groups of bacteria, proving the applicability of nose-type systems to this class of tasks. In this case, PCA is able to reduce six-dimensional data to only two dimensions, while preserving class separability in the resulting data.

An even more important feature of PCA is its ability to minimize sensor cross-correlation, which improves the accuracy of the PR algorithms. Radiation sensors normally are not selective to only one type (or dose rate) of radiation exposure. In practice, exposure of the radiation measuring system to even a single source will cause excitation of all sensors to various degrees. In general, PCA

Table 5.2
Abbreviations and References of Classification Techniques

MLR	Multiple linear regression [8]
PLS	Partial least squares [32]
PCA	Principal component analysis [37]
CA	Cluster analysis [37]
DFA	Discriminant function analysis [37]
PCR	Principal component regression [32]
SOM	Self-organizing map [32]
MLP	Multilayer perception [32]
PNN	Polynomial neural network [8]
RBF	Radial basis function [32]
LVQ	Learning vector quantization [8]
FIS	Fuzzy inference systems [8]
FNN	Fuzzy neural network [8]
FCM	Fuzzy c-means [8]
ART	Adaptive resonant theory [8]
Fuzzy ARTMAP	Fuzzy means an analog input may be presented to the classifier as opposed to a binary input with ARTMAP, and MAP represents the fact that the algorithm undergoes supervised training [33]
GA	Genetic algorithm [8]
NFS	Nero-Fussy Systems [8]
ANN	Artificial neural networks [37]
NN	Neural networks [8]
Ward's	A minimum-variance cluster analysis technique [8]
LDA	Linear discriminate analysis [8]

delivers good results for e-nose systems equipped with linear, cross-correlated sensors. PCA is also useful when the system contains considerably more sensors than necessary to detect the type or dose rate of interest in a particular application. In these cases, PCA will simply reduce data dimensionality. However, PCA will produce poor results if the sensors are strongly nonlinear or if sensor data is significantly affected by sensor drift. However, sensor nonlinearity is relatively easy to counteract through sensor calibration. A more serious problem is a sensor drift. A discussion on how to counteract sensor drift can be found in [8, 38].

5.4.3 Linear Discriminant Analysis

The main principle of linear discriminant analysis (LDA) is the maximization of class separability through linear transformation of the initial data coordinate system and a reduction of dimensionality [36]. LDA is not as popular as PCA, but it has proven its applicability in many e-nose systems. For example, the LDA algorithm was used by the National Aeronautic Space Administration (NASA) for space applications [39]. NASA was evaluating the ability to monitor air contaminants in a closed environment, such as the Space Shuttle or International Space Station (ISS). Two different e-nose systems, consisting of 7 and 38 MOS sensors, were used to detect volatile organic contaminants, notify of impending electrical fire by monitoring low-concentration chemical vapors emitted by overheating wires, and monitor hypergolic propellant contaminants in the airlock of the Shuttle and ISS. In all cases, LDA was able to distinguish between the various vapors with $\geq 90\%$ success rate.

LDA is very similar to PCA, except that PCA selects a new coordinate system to maximize variance. LDA instead selects a new coordinate system to maximize distance between class central points. This, in turn, increases the probability of proper classification. A detailed discussion of the LDA algorithm is presented in Section 3.8.2 of [36].

It is important to note that when the number of classes is greater than two a multiple discriminant analysis is employed, however it is based on the same principles as LDA and is often referred as LDA. Although LDA ensures the best separability of classes, it is not a classifier in itself. A classifier built on top of LDA typically uses a distance criterion to assign samples to one of the classes, using either the minimum distance technique or a more complex one (e.g., considering standard deviation of the classes). LDA is one of the simplest and most efficient PR algorithms available, yet it delivers good results in many e-nose applications. For very complex patterns, however, LDA may deliver significantly poorer results than more complex algorithms, specifically those based on neural networks (NN). NN-based PR algorithms require considerably more expert knowledge to optimally choose training parameters.

A comprehensive description of PR methods and concepts, with examples in real-life applications (e.g., engineering, medicine, economy, geology), can be found in [40].

5.4.4 Validation

In a supervised classification system, once the training is completed it is required to test the classification algorithm's performance. Methods for validation include the holdout method, random subsampling, k-fold cross-validation, leave-one-out validation, and three-way splits [30]. The number of available data sets determines the choice of the method. The holdout method splits the data into two data subsets: the training set and the test set. The holdout method may not be the preferred choice, if the number of data sets is limited. The k-fold cross-validation splits the data by randomly selecting a set of test samples; the algorithm is trained and tested with these samples. From the original data set, a different set of test samples are taken and the algorithm is retrained and tested. Each iteration of this procedure will result in error estimation. The average of all estimates is calculated for the final result. For *k*-fold validation, the data set is split as follows: for each k experiments run, remove $1/k$ of the data set for testing and use remaining samples for testing. In leave-one-out for a data set with N examples, N experiments are run, and for each experiment one example is removed from the data set to be used as the set. The three-way data split divides the data set into three sets: training set, evaluation set, and test set. The training set is used to train the classifier; the evaluation set is used to test the new classification system, but it may also be part of the training as it may be used to adjust the algorithm parameters. The test set is the true unknown data for testing the system [30, 32].

5.4.4.1 Targeted Applications of Radiation Nose Systems

One of the potential applications of the radiation nose systems is clearly as a tool for security purposes, the prevention of terrorist attacks, or as a part of army ammunition. There are numerous cases, where the special forces or the army have to act in an unknown area or respond to an emergency without having any information regarding the presence of radiation and what level and type. A small pocket size radiation nose system, capable of mixed radiation dosimetry, with real-time alarm system could be very important in a situation like this. Section 5.5 gives an example of hardware and software of such a system.

5.5 Radiation Nose Using a Compaq iPAQ Designed for Resistive Sensors

A datalogger consisting of a Compaq iPAQ and a microconverter was developed, and its block diagram is shown in Figure 5.9. The program that logs the data was developed on a personal computer (PC) using LabVIEW and compiled for the pocket PC operating system using LabVIEW's PDA module. An Analog Devices microconverter ADuC812s that digitizes the analog data is connected to the iPAQ using a serial port. The ADuC812s incorporates a 12-bit ADC. The developed system may be used for recording data from any sensors or sensor systems that output an analog voltage of between 0V and 5V.

The software developed for the iPAQ was designed to allow for different communications settings. Data may be displayed as raw data as it was received

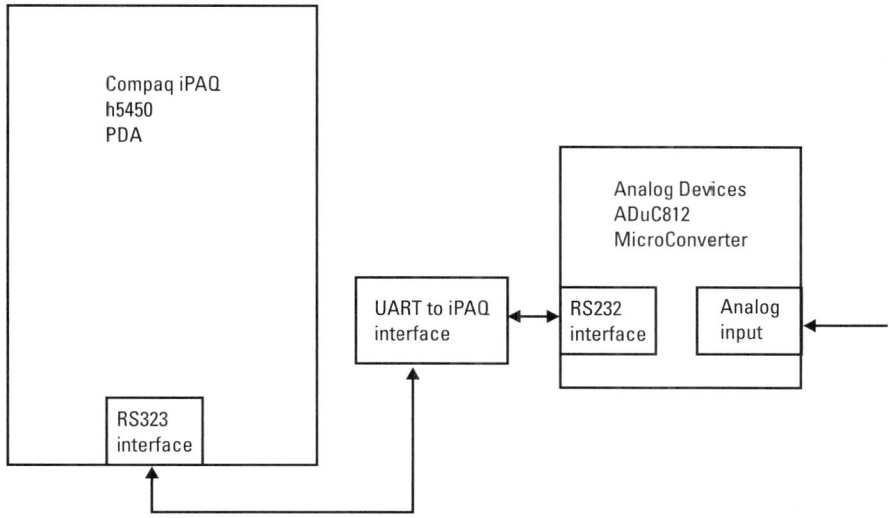

Figure 5.9 Block diagram of data logging system.

from the microconverter or it may be displayed as a graph of data against the number of samples recorded.

The iPAQ has a serial port that may be used to serially communicate with other serial devices. The microconverter has a universal asynchronous receiver/transmitter (UART) interface, and a serial transceiver chip is required to convert the UART's CMOS logic levels up to RS232 signal levels in order to allow it to successfully communicate with a computer or the iPAQ. A MAXIM MAX203 RS232 transceiver chip was used to accomplish this.

Both the microconverter and the iPAQ are portable and are small enough to fit in the palm of the hand. The iPAQ is a handheld device with a 400-MHz ARM processor. An extra 256 MB of flash memory was installed in the secure digital expansion slot to store the logged information.

The user interface presents the user with dialogs to allow parameters to be changed depending on the applications. Options that may be changed are serial port settings and sampling methods. The sampling methods available to the user through the interface are those that are implemented in the microconverter. These sampling methods include single sample, continuous sample, burst sample, and timed sample. The sampled data that is acquired by the iPAQ may be saved as a comma separated variable (CSV) file with a unique filename generated by the current date and time. This CSV file can be easily downloaded to a PC and opened in Microsoft Excel for further manipulation and analysis.

Figure 5.10 shows an image of the developed prototype handheld device that may be used for interfacing resistive-based sensor arrays for real-time radiation monitoring. The system boasts a modular nose head design, allowing different arrays to be connected quickly. This allows the device to be configured for a specific task if required. The system can interface to an array with up to eight sensors whose analog signals are multiplexed en route to the ADC. The PDA is capable of storing the data for further analysis; it also has built-in software to display the acquired data in real time. The PDA has the capability to run PR algorithms to execute classification of radiation types and doses. Even hybrid sensor arrays could be interfaced to the system, allowing for a dual-functioning device, encompassing e-nose and radiation nose functionality. For example, half the array could make up gas sensors for the detection of the explosives, and the other half could make up radiation sensors, allowing the overall system to be used for security and military applications.

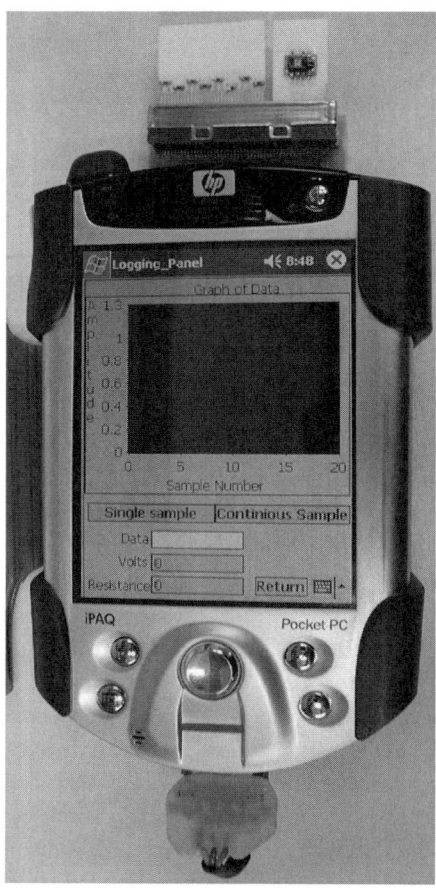

Figure 5.10 Prototype radiation nose system based on iPAQ.

5.6 Portable Real-Time Gamma Radiation Dosimetry System Using MgO and CeO$_2$ Thick Film Capacitors

In aggressive industrial processes and hard-to-reach places, such as power plants and radioactive waste storage places, it is extremely important to constantly monitor the level of radiation and to give corresponding alert signal in case of emergency. The solution to this problem could be offered via a portable real-time radiation monitoring system, in which case the risk of personnel exposure is minimized. An example of such a system is presented next, where thick film MgO and CeO$_2$ capacitors with Ag interdigitated electrodes were used as prototype radiation sensors. A ^{137}Cs (0.662 MeV) disk-type source (provided by AEA Technology QSA GmbH as a standard reference gamma radiation source) was used to expose the sample to γ-radiation.

5.6.1 Change of Capacitance with Dose

Figure 5.11 shows the change in the value of capacitance with radiation dose for a MgO thick film capacitor with interdigitated electrodes [41]. A monotonic increase in the values of capacitance from 1.493 pF to 1.537 pF was recorded as a result of 32.55-mGy γ-dose. Figure 5.12 shows the change in the value of capacitance with radiation dose for the counterpart CeO$_2$ samples. They were less susceptible to gamma radiation, as the value of capacitance increased from

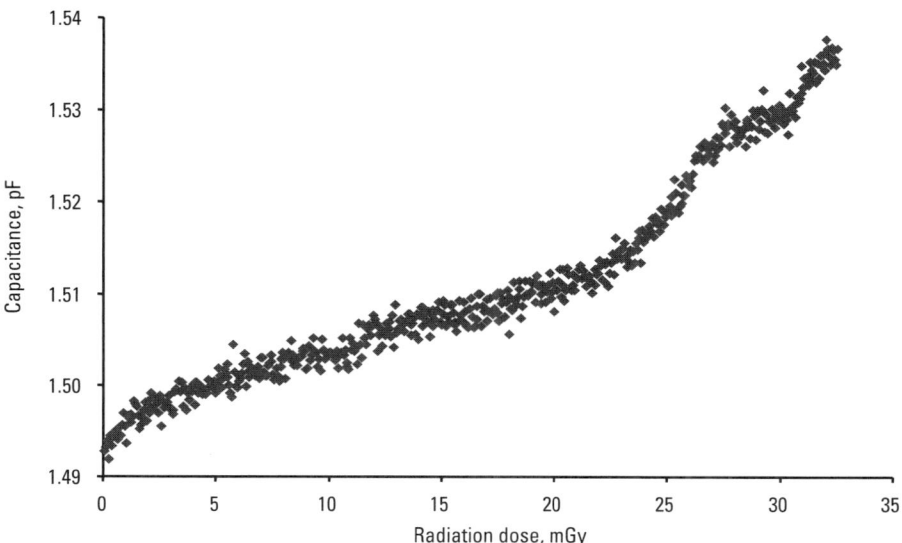

Figure 5.11 Change in the value of capacitance with radiation dose for a MgO thick film capacitor. (*From:* [41]. © 2005 IEEE. Reprinted with permission.)

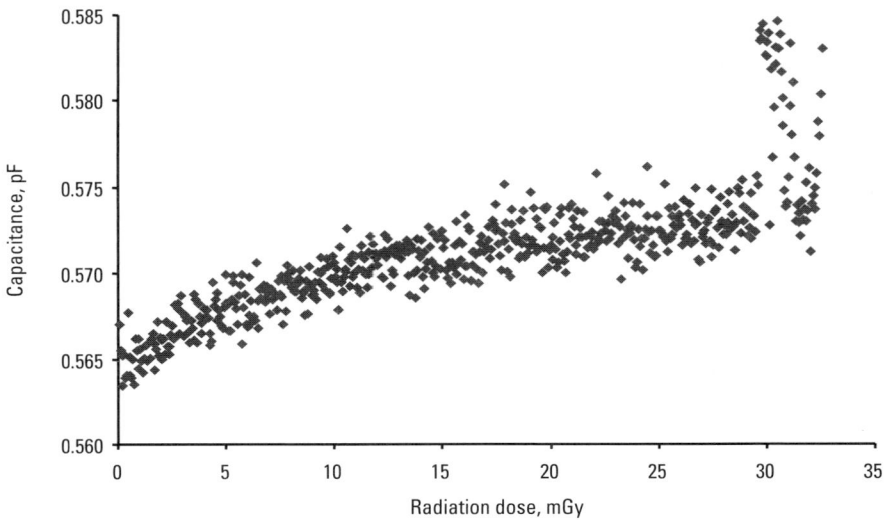

Figure 5.12 Change in the value of capacitance with radiation dose for a CeO_2 thick film capacitor. (*From:* [41]. © 2005 IEEE. Reprinted with permission.)

0.567 pF for the nonirradiated sample to 0.583 pF for irradiated with the dose of 32.55 mGy. Moreover, the response of the MgO sensor was monotonic over the whole range, which is preferable for dosimetry applications.

5.6.2 Circuit

The circuit design required a system capable of measuring capacitances changes with high precision. For this, the AD7746 (provided by Analog Devices, Inc., http://www.analog.com) 24-bit capacitance to digital converter was chosen for its inherent high-resolution architecture. The AD7746 allows 19-bit effective resolution at a 16.6-Hz data rate, high linearity (±0.01%), and high accuracy (±2 fF factory calibrated). The AD7746 capacitance input range is ±4 pF (changing), while it can accept up to 17 pF absolute capacitance (not changing), which is compensated by an on-chip digital to capacitance converter. In order to control the AD7746, a suitable processor was required. The ADuC831 was chosen and interfaced to the AD7746 by use of the I^2C bus. The AD7746 can be fully controlled from the I^2C bus and also allows multiple AD7746s to be connected simultaneously [41]. Figure 5.13 shows a block diagram of the AD7746.

In Figure 5.14, a block diagram of the ADuC831 microprocessor is presented. The only functionality required from this tightly packed IC was that of the I^2C bus, interrupt line input port p3.2 and an output port p3.4 to show that the system is functional. Although this processor probably exceeds the

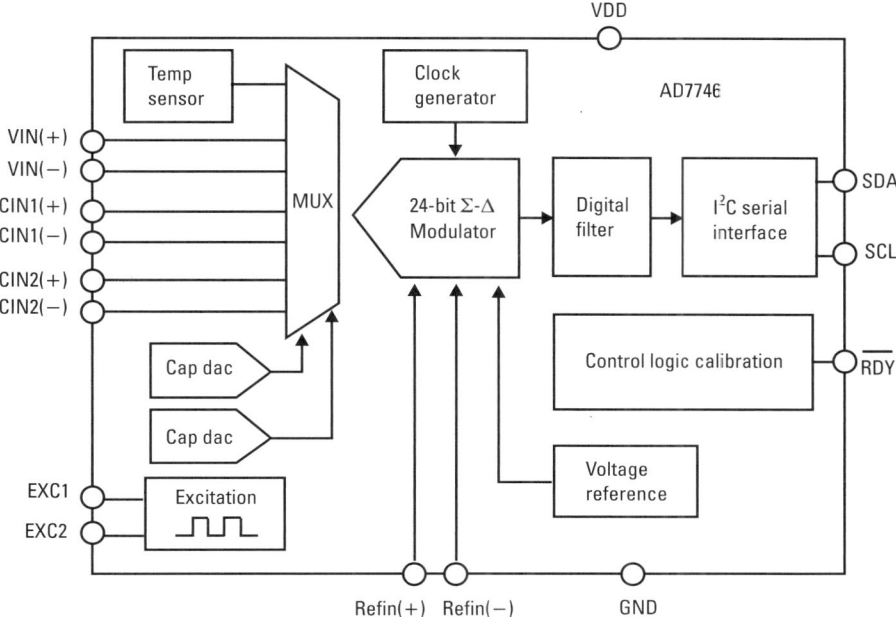

Figure 5.13 Functional diagram of the AD7746. (*From:* [42]. © 2005 Analog Devices, Inc. Reprinted with permission.)

requirements of this application, its ease of use with regard to I^2C routines and RS232 programming made it a suitable candidate for the initial test system [41].

The system can run on a direct 5-V or 3-V supply. An ADP3303 power supply regulator was used to power the microconverter (shown in Figure 5.15) to reduce power consumption and minimize noise in the capacitance circuit. The ADM202 simply converts the UART signals from the ADuC831 to the correct voltage levels, TTL to inverted CMOS, so that the system can be interfaced to a PC. The full schematic for the prototype system can be seen Figure 5.16.

5.6.3 Software

Some initial assembly code was written and downloaded to the ADuC831 internal EEPROM. This allowed for the system to be fully controllable with regard to time and resolution from the host-based PC system. Software was developed in LabVIEW that controls the system and displays the required data for further analysis. References and code samples for the internal workings of the ADuC831 with regard to I^2C bus or serial port communication can be found in the AD7745.46 datasheet [42].

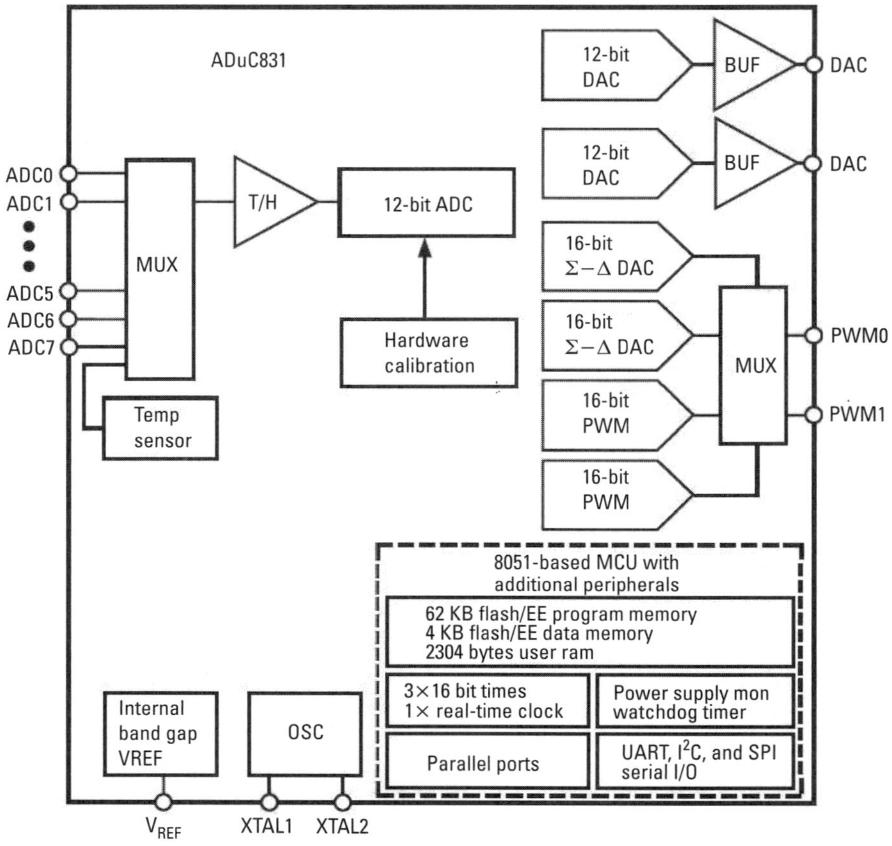

Figure 5.14 Functional diagram of ADuC831. (*From:* [42]. © 2005 Analog Devices, Inc. Reprinted with permission.)

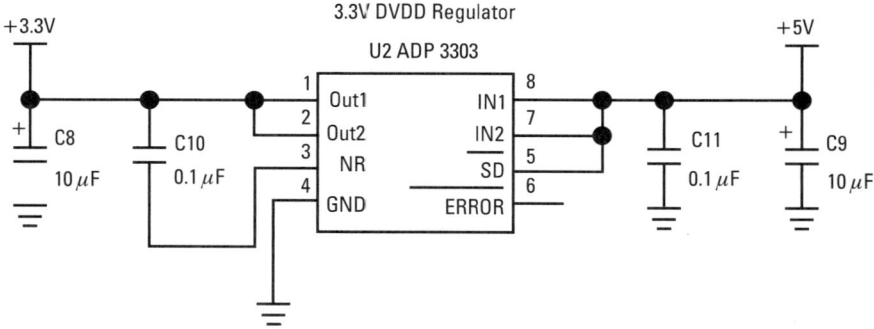

Figure 5.15 Regulated power supply for circuit. (*From:* [42]. © 2005 Analog Devices, Inc. Reprinted with permission.)

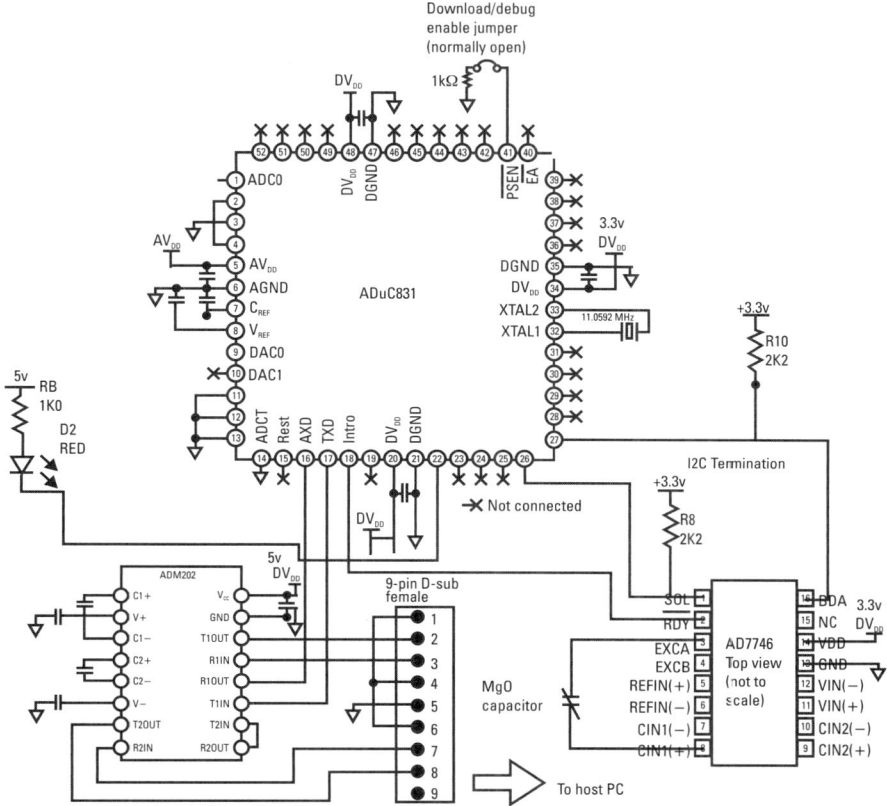

Figure 5.16 Full prototype circuit for testing low-range capacitance changes in irradiated MgO and CeO$_2$ capacitors. (*From:* [42]. © 2005 Analog Devices, Inc. Reprinted with permission.)

5.6.3.1 Assembly Code

Figure 5.17 represents the assembly code in graphical form [41]. The code begins by setting the parameters for the serial port baud settings (9600, 8, n, 1) and initializing the I^{2}C bus. The system interrupts are then set to be edge triggered and enabled. The parameters for the slave address and the configuration mode are stored. The I^{2}C bus register is set so that the ADuC831 is in master mode, the SDATA and SCLOCK lines are high, and the status bits are cleared.

The software embedded in the microconverter polls the serial buffer in anticipation of receiving the control signal either 1 or 2 to initiate a data acquisition routine or asleep routine, respectively. The data acquisition routine starts the gathering capacitance value routine by first acquiring the slave device (AD7746) on the I^{2}C bus, setting it to single conversion mode with capacitor

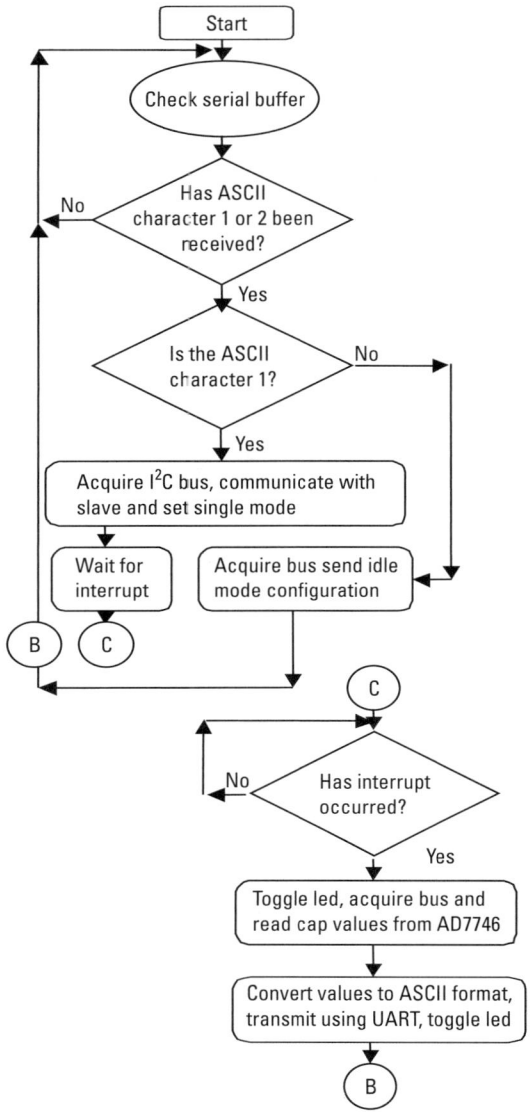

Figure 5.17 Graphical representation of assembly code. (*From:* [41]. © 2005 IEEE. Reprinted with permission.)

inputs on the C+ and C− terminals, at a rate of 16 Hz (best noise performance, according to the AD7745.46 datasheet [42]). The AD7746 then takes a capacitance reading and sets its ready signal low. The 24-bit capacitance value is then stored internally on the AD7746 in registers 0x01, 0x02, and 0x03. On the ready signal going low, an interrupt is generated causing the ADuC831 to flash the LED connected to pin 3.4. It then acquires the I²C bus and accesses the capacitor storage registers. The values are then stored internally in the ADuC831's RAM before being converted to ASCII code and transmitted through the UART to the host PC system. If the ASCII character 2 is received, the ADuC831 again acquires the I²C bus and sends the IDLE mode value to the configuration register of the AD7746. The assembly code was compiled using the ASM51 assembler and downloaded using the serial programmer port of the ADuC831 [41].

5.6.3.2 LabView Program

LabView software was chosen as the simplest way of collecting data from an external source and collating it into viewable charts on a PC-based system. The user sets the time base in milliseconds between samples and presses the go button. Figure 5.18 displays the methodology behind the LabView [41].

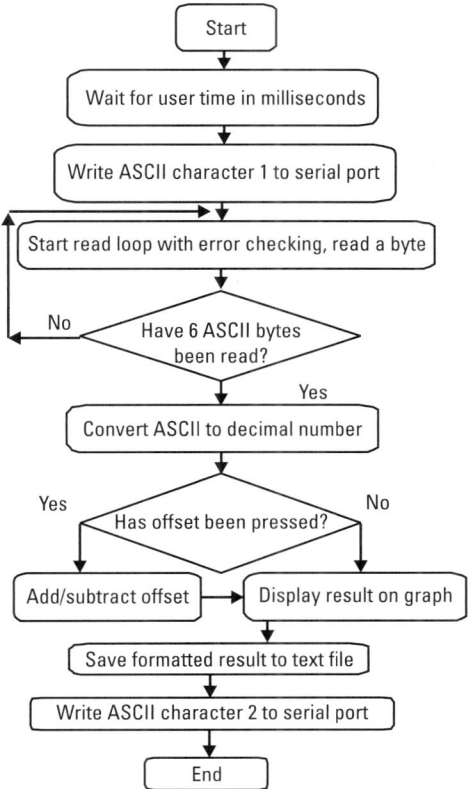

Figure 5.18 Graphical representation of LabView program. (*From:* [41]. © 2005 IEEE. Reprinted with permission.)

It can be seen that the system outlined here provides a user-friendly method of collecting data from the radiation sensor under test and creating viewable charts on a PC-based system. This portable real-time dosimetry system can be used in a wide range of applications, such as security tasks, environmental monitoring, and nuclear waste control, where personnel exposure to radiation can be minimized. This system has the capability of being networked using wireless technology for remote radiation monitoring.

References

[1] Arshak, K., and O. Korostynska, "Gamma Radiation Nose System Based on In_2O_3/SiO Thick Film Sensors," *IEEE Sensors,* in press, 2006, available online in "Forthcoming Publications."

[2] Shuduo, W., "Polyvinylidene Fluoride Film Sensors and Applications," *7th Intl. Symp. on Proc. Electnets,* Berlin, Germany, September 25–27, 1991, pp. 923–928.

[3] Torrisi, L., et al., "Radiation Effects of keV-MeV Ion Irradiated PVDF," *Nuclear Instruments and Methods in Physics Research, Section A: Accelerators, Spectrometers, Detectors and Associated Equipment,* Vol. 382, No. 1–2, 1996, pp. 361–364.

[4] Hongbing, S., et al., "Radiation Effects of Poly (Vinylidene Fluoride) (PVDF) (I)—Mechanism of Accelerating Damage of Crystal in High Dose Zone," *Science in China, Series A: Mathematics, Physics, Astronomy,* Vol. 41, No. 4, 1998, pp. 438–442.

[5] Atanassova, E., et al., "Influence of γ-Radiation on Thin Ta_2O_5-Si Structures," *Microelectronics Journal,* Vol. 32, No. 7, 2001, pp. 553–562.

[6] Arshak, K., and O. Korostynska, "Influence of Gamma Radiation on the Electrical Properties of MnO and MnO/TeO_2 Thin Films," *Annalen der Physik,* Vol. 13, No. 1–2, 2004, pp. 87–89.

[7] Arshak, A., et al., "High Dose Optical and Electrical Sensor Dosimeter Using Cobalt Phthalocyanine (CoPc) Thick Film," *Proc. IMAPS,* 2002, pp. 766–771.

[8] Pearce, T. C., et al., *Handbook of Machine Olfaction (Electronic Nose Technology),* New York: Wiley-VCH Verlag Gmbh, 2003.

[9] Gardner, J. W., E. L. Hines, and M. Wilkinson, "Application of Artificial Neural Networks to an Electronic Olfactory System," *Measurement Science & Technology,* Vol. 1, No. 5, 1990, pp. 446–451.

[10] Hartman, J. D., "A Possible Objective Method for the Rapid Estimation of Flavors in Vegetables," *Proceedings of the American Society for Horticultural Science,* Vol. 64, 1954, pp. 335–342.

[11] Muñoz, B. C., G. Steinthal, and S. Sunshine, "Conductive Polymer-Carbon Black Composites-Based Sensor Arrays for Use in an Electronic Nose," *Sensor Review,* Vol. 19, No. 4, 1999, pp. 300–305.

[12] O'Connell, M., et al., "A Practical Approach for Fish Freshness Determinations Using a Portable Electronic Nose," *Sensors and Actuators B: Chemical,* Vol. 80, No. 2, 2001, pp. 149–154.

[13] Brezmes, J., et al., "Correlation Between Electronic Nose Signals and Fruit Quality Indicators on Shelf-Life Measurements with Pinklady Apples," *Sensors and Actuators B: Chemical,* Vol. 80, No. 1, 2001, pp. 41–50.

[14] Nagle, H. T., R. Gutierrez-Osuna, and S. S. Schiffman, "The How and Why of Electronic Noses," *IEEE Spectrum,* Vol. 35, No. 9, 1998, pp. 22–31.

[15] Llobet, E., et al., "Fuzzy ARTMAP Based Electronic Nose Data Analysis," *Sensors and Actuators B: Chemical,* Vol. 61, No. 1–3, 1999, pp. 183–190.

[16] Miekisch, W., J. K. Schubert, and G. F. E. Noeldge-Schomburg, "Diagnostic Potential of Breath Analysis—Focus on Volatile Organic Compounds," *Clinica Chimica Acta,* Vol. 347, No. 1–2, 2004, pp. 25–39.

[17] Di Natale, C., et al., "Lung Cancer Identification by the Analysis of Breath by Means of an Array of Non-Selective Gas Sensors," *Biosensors and Bioelectronics,* Vol. 18, No. 10, 2003, pp. 1209-1218.

[18] Phillips, M., et al., "Detection of Lung Cancer with Volatile Markers in the Breath," *Chest,* Vol. 123, No. 6, 2003, pp. 2115–2123.

[19] Phillips, M., et al., "Volatile Markers of Breast Cancer in the Breath," *Breast Journal,* Vol. 9, No. 3, 2003, pp. 184–191.

[20] Gardner, J. W., H. W. Shin, and E. L. Hines, "An Electronic Nose System to Diagnose Illness," *Sensors and Actuators B: Chemical,* Vol. 70, No. 1–3, 2000, pp. 19–24.

[21] Holmberg, M., et al., "Identification of Paper Quality Using a Hybrid Electronic Nose," *Sensors and Actuators B: Chemical,* Vol. 27, No. 1–3, 1995, pp. 246–249.

[22] Harwood, D., "Something in the Air [Electronic Nose]," *IEE Review,* Vol. 47, No. 1, 2001, pp. 10–14.

[23] Bourgeois, W., and R. M. Stuetz, "Use of a Chemical Sensor Array for Detecting Pollutants in Domestic Wastewater," *Water Research,* Vol. 36, No. 18, 2002, pp. 4505–4512.

[24] Gardner, J. W., et al., "An Electronic Nose System for Monitoring the Quality of Potable Water," *Sensors and Actuators B: Chemical,* Vol. 69, No. 3, 2000, pp. 336–341.

[25] Moore, D. S., "Instrumentation for Trace Detection of High Explosives," *Review of Scientific Instruments,* Vol. 75, No. 8, 2004, pp. 2499–2512.

[26] Yinon, J., "Detection of Explosives by Electronic Noses," *Analytical Chemistry,* Vol. 75, No. 5, 2003, pp. 99A–105A.

[27] Arshak, K., et al., "A Review of Digital Data Acquisition Hardware and Software for a Portable Electronic Nose," *Sensor Review,* Vol. 23, No. 4, 2003, pp. 332–344.

[28] Arshak, K., et al., "A Review of Gas Sensors Employed in Electronic Nose Applications," *Sensor Review,* Vol. 24, No. 2, 2004, pp. 181–198.

[29] Horowitz, Y. S., "Theory of Thermoluminescence Gamma Dose Response: The Unified Interaction Model," *Nuclear Instruments and Methods in Physics Research Section B: Beam Interactions with Materials and Atoms,* Vol. 184, No. 1–2, 2001, pp. 68–84.

[30] Strathmann, S., and F. Dieterle, "Introduction to Data Evaluation for Sensor Systems," in *Nose Lecture Series: Data Evaluation, Nose II Network,* S. Strathmann, M. Harbeck, and U. Weimar, (eds.), 2002, pp. 1–36; *Proc. of 3rd European Short Course of the NOSE II NETWORK,* Alpbach, Austria, March 21–26, 2004.

[31] Pearce, T. C., "Computational Parallels Between the Biological Olfactory Pathway and Its Analogue the Electronic Nose: Part II. Sensor-Based Machine Olfaction," *Biosystems,* Vol. 41, No. 2, 1997, pp. 69–90.

[32] Gutierrez-Osuna, R., "Pattern Analysis for Machine Olfaction: A Review," *Sensors Journal,* Vol. 2, No. 3, 2002, pp. 189–202.

[33] Carpenter, G. A., et al., "Fuzzy ARTMAP: A Neural Network Architecture for Incremental Supervised Learning of Analog Multidimensional Maps," *IEEE Trans. on Neural Networks,* Vol. 3, No. 5, 1992, pp. 698–713.

[34] Perera, A., et al., "A Portable Electronic Nose Based on Embedded PC Technology and GNU/Linux: Hardware, Software and Applications," *IEEE Sensors Journal,* Vol. 2, No. 3, 2002, pp. 235–246.

[35] Pardo, A., et al., "Methods for Sensors Selection in Pattern Recognition," *Proc. 7th Int. Symposium on Olfaction and Electronic Noses, Inst. of Physics Publishing,* July 2000, pp. 83–88.

[36] Duda, R. O., P. E. Hart, and D. G. Stork, *Pattern Classification,* New York: John Wiley & Sons, 2001.

[37] Hines, E. L., E. Llobet, and J. W. Gardner, "Electronic Noses: A Review of Signal Processing Techniques," *IEE Proceedings Circuits, Devices and Systems [see also IEE Proceedings G—Circuits, Devices and Systems],* Vol. 146, No. 6, 1999, pp. 297–310.

[38] Holmberg, M., et al., "Drift Counteraction in Odour Recognition Applications: Lifelong Calibration Method," *Sensors and Actuators B: Chemical,* Vol. 42, No. 3, 1997, pp. 185–194.

[39] Young, R. C., et al., "Electronic Nose for Space Program Applications," *Sensors and Actuators B: Chemical,* Vol. 93, No. 1–3, 2003, pp. 7–16.

[40] de Sa Marques, J. P., *Pattern Recognition: Concepts, Methods, and Applications,* Berlin, Germany: Springer-Verlag, 2001.

[41] Arshak, K., et al., "Portable Real-Time Gamma Radiation Dosimetry System Using MgO and CeO_2 Thick Film Capacitors," *Proc. IEEE Int. Conf. on Sensing Technology (ICST),* November 21, 2005, pp. 137–142.

[42] Analog Devices, Microconverters, http://www.analog.com/microconverters, 2005.

6

Conclusions and Future Trends

6.1 Necessity of Radiation Dosimetry

Occupational exposure to ionizing radiation can occur in a range of industries, medical institutions, educational and research establishments, and nuclear fuel cycle facilities. Adequate dosimetry is essential for the safe and acceptable use of radiation, radioactive materials, and nuclear energy. Additional interest to the area of radiation detector development has grown considerably due to all aspects related to homeland security. Considering numerous nuclear accidents (briefly reviewed in Section 2.6 of Chapter 2) and the possibility of facing such issues as lost radiation sources, transportation and storage of nuclear waste, radiological terrorism, and the possibility of nuclear weapons being used in a war, the members of the public should be encouraged to use personal electronic dosimeters, which could provide real-time alert if the safe limit of the external dose is exceeded. For a long time, radiation dosimetry was a necessity only for the nuclear specialists, military forces, and medical personnel. Radiation is currently used in many areas and, unfortunately, quite often affects civilians, ordinary workers, and inhabitants. The use of personal real-time radiation detectors is especially crucial for first responders (to alert to radioactive threats); customs and border patrols; security officers in banks, government laboratories, and medical and research facilities; and police and fire departments.

6.2 The Choice of Detector

The choice of a particular detector type for an application depends on the radiation energy range of interest, the application's resolution, the efficiency requirements, the detector's performance, timing, environmental suitability, and cost. Due to the large number of different applications, there are many kinds of radiation sensors. Various materials, geometric arrangements, and diverse physical-detection techniques are used, which are discussed in Chapter 2. Detection of radiation is based on the fact that electrical, optical, and structural properties of the materials undergo changes under the influence of ionizing radiation. The effect of irradiating an electronic material and the consequent degradation in performance of devices made from such a material can follow a number of routes. The final result depends on the type and energy of incident radiation, the type of materials, their particular contribution to the device function, and the physical principles upon which the function of the device is based [1]. Deep understanding of the physical properties of the materials under the influence of radiation exposure is vital for the effective design of dosimeter devices. This book covers the basic theory of radiation physics and radiation sensing mechanisms of various materials. Up-to-date data on commercially available personal radiation dosimeters is presented, along with a list of companies that manufacture and distribute various radiation-detection equipment. This information could serve as a reference guide for engineers, designers, and scientists working in this area.

6.3 Standards and Requirements

There is also a sizeable variety of detectors for radiation monitoring; they are strictly examined to correspond with the standards and requirements (see Section 2.7.2). For example, the European Radiation Dosimetry Group (EURADOS) aims to assist in the process of harmonization of individual monitoring as part of the protection of occupationally exposed workers in 26 European countries [2]. The objective of their tasks, related to the assessment of doses in either internal or external exposures, is the integration of dosimetric methods—investigating how the results from personal dosimeters for external radiation and workplace monitoring and from monitoring for internal exposure can be combined into a complete and consistent system of individual monitoring. It is important to investigate how these different methods can be correlated so that the numerical dose values can be added to result in a total effective dose for the worker. There is a considerable concern within regulatory bodies and approved internal dosimetric services regarding the need for standardization for the evaluation methods of internal exposures. In this way, national and international intercomparisons have been organized for the purpose of checking not only the performance of in vivo and in vitro laboratories, but also the methodology used by services to assess the effective doses, taking into account ICRP recommendations and national regulations [2]. The objective is always to validate measurement procedures and dosimetric tools to guarantee the reliability of calculated doses.

6.4 Future Trends

The future of dosimetry runs in two parallel directions that depend on application (e.g., high-precision expensive dosimetry for nuclear, medical, and space purposes) and alternative wide-range real-time dosimetry with an affordable device price even for the personal use of civilians. For example, this could be achieved via the use of cost-effective metal oxides and metal phthalocyanines thin/thick films, which are discussed in detail in Chapter 4. These devices cannot compete, for instance, with Si-based diodes (see Section 2.7.10), RADFETs (see Section 2.7.8), or CZT materials (see Section 2.7.11.4) in terms of their superior spatial resolution and efficiency; however, they are favorable for real-time safeguard monitoring to insure that the radiation exposure is within the safe limits. A few devices based on these materials are presented in Chapter 5, and currently a wireless personal dosimetry system is being developed.

The progress made in the area of radiation dosimetry became possible due to the achievements made in such areas as semiconductor industry, material processing and characterization, and novel manufacturing technologies that have emerged recently. This enabled considerable miniaturization of the devices and their wireless/remote operation, which in a number of cases is vital. Only close collaboration between scientists, researchers, engineers, medical doctors, and so forth would facilitate the development of novel approaches to further the design of both advanced high-precision and affordable radiation sensors.

References

[1] Holmes-Siedle, A. G., and L. Adams, *Handbook of Radiation Effects*, New York: Oxford University Press, 1993.

[2] Lopez Ponte, M. A., et al., "Individual Monitoring for Internal Exposure in Europe and the Integration of Dosimetric Data," *Radiation Protection Dosimetry*, Vol. 112, No. 1, 2004, pp. 69–119.

Acronyms

ABS	Acrylonitrile/butadiene/styrene
ADC	Analog to digital converter
ALARA	As low as reasonably achievable
ANN	Artificial neural networks
ARS	Agricultural Research Service
ART	Adaptive resonant theory
ASCII	American standard code for information interchange
ASTAR	Database for stopping powers and ranges for helium ions
BEIR Committee	National Academy of Sciences Committee on the Biological Effects of Ionizing Radiation
BGO	Bismuth germanate $Bi_4Ge_3O_{12}$
BNFL	British Nuclear Fuels
CA	Cluster analysis
CCD	Charge-coupled device
CCs	Competitive centers
CMOS	complementary metal oxide semiconductor
CPU	Central processing unit

CSV	Comma separated variable
CT	Computed tomography
CVD	Chemical vapor deposition
Cz-Si	Czochralski silicon
CZT	$Cd_{1-x}Zn_xTe$
DFA	Discriminant function analysis
DNA	Deoxyribonucleic acid
EEPROM	Electrically erasable programmable read-only memory
e-nose	Electronic nose
EPDM	Ethylene-propylene diene monomer
ESR	Electron spin resonance
ESTAR	Database for stopping powers and ranges for electrons
ETFE	Ethylene-tetrafluoroethylene
EURADOS	European Radiation Dosimetry Group
fcc	Face centered cubic
FCM	Fuzzy c-means
FEP	Tetrafluoroethylene-hexa-fluoropropylene
FET	Field-effect transistor
FIS	Fuzzy inference systems
FNN	Fuzzy neural network
FWHM	Full width at half maximum
Fz-Si	Float zone silicon
GA	Genetic algorithms
GEM	General effective media
GM	Geiger-Mueller
HBM	Horizontal Bridgman method
HPBM	High-pressure Bridgman method

IAEA	International Atomic Energy Agency
ICRP	International Commission on Radiological Protection
ICRU	International Commission on Radiation Units and Measurements
IEC	International Electrotechnical Commission
INES	International Nuclear Event Scale
ISO	International Organization for Standardization
ISS	International Space Station
LCD	Liquid crystal display
LCs	Luminescent centers
LDA	Linear discriminate analysis
LED	Light-emitting diode
LET	Linear energy transfer
LLD	Lower limit of detection
LLNL	Lawrence Livermore National Laboratory
LNT	Linear nonthreshold
LVQ	Learning vector quantization
LWR	Light water reactor
MBE	Molecular beam epitaxy
MDA	Minimum detectable amount
MDD	Minimum detectable dose
MePc	Metal-substituted phthalocyanines
MLP	Multilayer perception
MLR	Multiple linear regression
MOS	Metal-oxide-semiconductor
MOSFET	Metal oxide silicon field effect transistor
MRI	Magnetic resonance imaging
NASA	National Aeronautic Space Administration

NCRP	National Council on Radiation Protection and Measurement
NEA	Nuclear Energy Agency of the Organization for Economic Cooperation and Development
NFS	Nero-fuzzy systems
NIST	National Institute of Standards and Technology
NN	Neural networks
OSL	Optically stimulated luminescence
PC	Personal computer
PCA	Principal component analysis
PCR	Principal component regression
PCTFE	Polychlorotrifluoroethylene
PFA	Tetrafluoroethylene-per-fluoromethoxyethylene
PIPS	Passivated ion-implanted silicon
PLS	Partial least squares
PMMA	Polymethylmethacrylate
PMT	Photomultiplier tube
PNN	Polynomial neural network
PR	Pattern recognition
PSTAR	Database for stopping powers and ranges for protons
PTFE	Tetrafluoroethylene
PVC	Polyvinylchloride
PVD	Physical vapor deposition
PVDC	Polyvinylidene chloride
PVDF	Polyvinylidene fluoride
RADFET	Radiation-sensing field-effect transistor
RBE	Relative biological effectiveness
RBF	Radial basis function

RDX	1,3,5-trinitro-1,3,5-triazocyclohexane
RPL	Radiophotoluminescent
SAN	Styrene/acrylonitrile
SBF	Sequential backward selection
SEM	Scanning electron microscope
SFS	Sequential forward selection
SI	International System of Units
SOM	Self-organizing map
SPCLC	Space-charge-limited conduction
SSB	Silicon surface barrier
TCs	Trapping centers
THM	Traveling heater method
TLD	Thermoluminescent dosimeter
TMI-2	Three Mile Island Unit 2
TNT	2,4,6-trinitrotoluene
UART	Universal asynchronous receiver/transmitter
UNSCEAR	United Nations Scientific Committee on the Effects of Atomic Radiation
UV	ultraviolet
UV-VIS	ultraviolet-visible
XRD	X-ray diffraction.

Appendix

Companies That Manufacture/Distribute Radiation Detection Equipment

Alpha Spectra, Inc.	http://www.alphaspectra.com
ANTECH Corporation	http://www.antech-inc.com
Arrow-Tech, Inc.	http://www.arrowtechinc.com
Atlantic Nuclear Corp.	http://atnuke.com
Berkeley Nucleonics Corp.	http://www.berkeleynucleonics.com
Canberra Industries	http://www.canberra.com
Far West Technology, Inc.	http://www.fwt.com
F&J Specialty Products, Inc.	http://www.fjspecialty.com
Femto-Tech, Inc.	http://www.femto-tech.com
Fluke Biomedical	http://www.flukebiomedical.com
Gamma Products, Inc.	http://www.gammaproducts.com
GEX Corporation	http://www.gexcorporation.com
Global Dosimetry Solutions, Inc.	http://www.globaldosimetry.com
Hopewell Designs, Inc.	http://www.hopewelldesigns.com
Images SI, Inc.	http://www.imagesco.com
J. L. Shepherd & Associates	http://www.jlshepherd.com
Laboratory Impex Systems LTD	http://www.lab-impex-systems.com
Lancs Industries	http://www.lancsindustries.com
Landauer, Inc.	http://landaueriii.com
Ludlum Measurements, Inc.	http://www.ludlums.com
SynOdys Group	http://www.synodys.com

MOHAWK Industrial and Nuclear Supply, nc.	http://www.mohawksafety.com
Nukepills.com	http://www.nukepills.com
Polimaster, Inc.	http://www.polimaster.com
ORTEC	http://www.ortec-online.com
Protean Instrument Corp.	http://www.proteaninstrument.com
Radiation Safety & Control Services, Inc.	http://www.radsafety.com
Radiation Safety Associates, Inc.	http://www.radpro.com
Radiation Sensors, LLC	http://www.radiationsensors.com
RITEC Ltd.	http://www.ritec.lv
RSO, Inc.	http://www.rsoinc.com
S. E. International, Inc.	http://www.seintl.com
SOLTEC Corp.	http://www.solteccorp.com
Spectrum Techniques	http://www.spectrumtechniques.com
Technical Associates	http://www.tech-associates.com
Teletrix Corporation	http://www.teletrix.com
Thermo Electron Corporation	http://www.thermo.com
XRF Corporation	http://www.xrfcorp.com

About the Authors

Khalil Arshak received a B.Sc. in physics from Basrah University, Iraq, in 1969; an M.Sc. in solid state physics from Salford University, United Kingdom, in 1979; and a Ph.D. (solid state electronics of thin films) and a D.Sc. (semiconductors) from Brunel University, United Kingdom, in 1986 and 1998, respectively. He joined the University of Limerick in 1986, where he leads the Microelectronic and Semiconductor Research Group. He has authored more than 250 research papers in the area of microelectronics and thin/thick film technology, FIB lithography and lithography process modeling, top surface imaging processes characterization, mixed oxide thin and thick film sensor development, and application-specific integrated circuit design.

Olga Korostynska received a B.Sc. and an M.Sc. from the National Technical University of Ukraine (KPI) in 1998 and 2000, respectively, in biomedical electronics. In 2003, she received a Ph.D. from the Department of Electronic and Computer Engineering, University of Limerick, Ireland, where she stayed as a postdoctoral research fellow. Her research interests are sensors development using thin and thick film technologies, material properties characterization, cost-effective gamma radiation dosimeters, and pressure sensors for medical device applications.

Arousian Arshak is the coauthor of Chapter 4. She received a B.Sc. in physics from Basrah University, Basrah, Iraq, in 1969; an M.Sc. in atomic physics from Salford University, Salford, United Kingdom, and a Ph.D. in optical lithography from the University of Limerick, Limerick, Ireland. Currently she is a lecturer in the Physics Department of the University of Limerick. Her research interests include optical, FIB, and soft lithography; modeling of silylation processes; mechanical and dielectric properties of thin films for circuits interconnects; and radiation damage in phthalocyanine thin/thick films for sensors applications.

Saleh M. Zleetni is the coauthor of Chapter 4. He received a Ph.D. in physics from the University of Limerick, Limerick, Ireland, in 2002. His research work was entitled Investigations and Feasibility Studies of Metal-Phthalocyanines for Radiation Dosimetry. Currently he is the head of the Research and Development Sector in National Authority of Information of Libya.

Colm Cunniffe is the coauthor of Chapter 5. He received a B.Tech. in information technology and telecommunications and a Ph.D. from the University of Limerick, Limerick, Ireland, in 2002 and 2006, respectively. He joined the Microelectronic and Semiconductor Research Group in the University of Limerick in 2002, where he is involved in research in the area of handheld electronic noses, embedded software, and pattern analysis, with applications in the food safety and medical diagnostics industries. Other research interests include characterization and test automation for sensors, sensor systems and design, and test of sensor sampling systems.

Index

Absorption
  coefficient μ, 33, 58, 74
  edge, 93, 96, 128, 149
  spectra, 64, 96, 119, 153
Activation energy, 95, 101, 103
Acute dose, 26, 27
Alanine, 80
ALARA principle, 31
Alpha (α) particles, 1, 12, 15, 22, 30, 56, 72
amorphous film, 92, 94
Angiography, 37
Annealing, 47, 65, 112, 136, 138, 142, 153
Annihilation, 107
Array, 7, 39, 49, 81, 139, 154, 159, 162–75
Arsenic sulfur (AChS) thin film, 118
Auger electron, 74

Backward diode, 131, 140, 142
Baseline manipulation, 168–70
  differential, 169
  fractional, 169, 170
  logarithm, 169, 170
  relative, 169, 170
Band gap, 7, 72, 75, 76, 80, 92, 97–98, 105, 120, 127, 132, 150
Beta (β) decay, 13, 73
Binding energy, 74, 105
Biological effects of radiation, 25
  half-life, 28
  tissue, 18, 22
Bismuth germanate (BGO), 57

Bq, 24, 41–45
Bremsstrahlung radiation, 12, 15, 18, 32, 73
Bruggeman symmetric medium, 139

Cadmium telluride (CdTe), 75, 80
Calcium fluoride (CaF), 63, 67
Calibration, 48–51, 61, 66, 79, 174
Cap de la Hague, 41, 42
Carbon black, 138, 139
CeO_2/Si backward diode, 142, 143
Cerium oxide thick film, 141, 142, 179, 180
Cermet, 139
Cesium iodide (CsI), 57, 60, 82, 83
Chain scission of polymers, 36, 108
Characteristic X-rays, 12
Charge carrier, 65, 72, 80, 95, 99, 121, 122, 135, 150
Charge transfer, 100, 121
Chemical dosimeters, 61
Chernobyl, 25, 39, 40, 44, 45
Chronic dose, 26, 27
Color center. *See* F-center, 6, 47, 64, 107, 112, 118, 135, 136
Compton effect, 58, 73–74, 105
Computed tomography, 2, 37, 79, 81
Conduction band, 5, 65, 76, 91–96, 104, 121, 128, 131, 150
  bulk limited, 99
  electronic, 102, 103
  hopping, 95, 102, 103
  impurity, 101, 102

Conduction band (continued)
  ionic, 101, 103
  space-charge-limited, 102, 134, 135
Conductivity σ, 5, 37, 47, 72, 91, 92, 96, 97, 99, 102, 105, 107, 108, 113, 116, 138–41, 151, 161, 164, 165
CoPc thick film, 6, 147–50, 162
Crosslinking of polymers, 35, 36, 108, 112
Crystalline structure, 5, 62–64, 67, 75, 92–95, 102, 103, 105, 118, 120, 124, 127, 142, 161
Curie (Ci), 24
CuPc, 6, 133–36, 145–51
Czochralski silicon, 77
CZT, 80, 81, 191

Density of states, 92, 127
Detection efficiency E_{ff}, 47
  sensitivity, 47, 49, 51, 53, 55, 63, 67, 68, 72, 150, 153
Detector resolution Γ, 51, 58, 73–75, 79–81, 191
Diamond dosimeter, 71, 72, 80
Dielectric
  constant, 6, 92
  response, 103
Diffusion, 65, 95, 117, 129, 131, 141, 153
Dose equivalent H, 22, 23, 29, 31, 63, 67, 79
  absorbed D, 2, 22–24, 29, 30, 34, 37, 51, 52, 58, 61, 62, 69, 72, 79, 84, 137, 153, 161
  ambient $H^*(d)$, 23
  directional $H'(d,\Omega)$, 23
  effective H_E, 29, 30, 43
  personal $H_p(d)$, 23, 24, 48, 53, 55
Drain current, 70, 71

Effective dose E, 2, 24, 28–30, 190
Electron, 1, 6, 11–15, 18, 23, 33, 34, 37, 39, 47, 51, 52, 56, 57, 62, 64, 65, 72–75, 80, 81, 93–95, 98, 101, 103, 105–8, 118, 121
  affinity, 99
  hole pair, 5, 14, 65, 72, 75, 91, 106, 107
Electronic nose, 163–65, 172
Energy-wave vector, 93
Esaki, 129, 131
EURADOS, 190
Evaporation chamber, 115

Excitation, 12, 15, 18, 22, 52, 57, 67, 72, 76, 95, 103, 105, 107, 112, 117, 118, 128, 173, 181
Exciton line theory, 96

Fading, 63, 68
Fano factor F, 52
F-center. *See also* Color center, 47, 64, 107, 108, 118
  feature
    extraction, 8, 169, 171
    selection, 169, 171, 172
Fermi level, 76, 81, 100, 103, 121, 128, 131
Film
  badge dosimeter, 66
  radiography, 49
Fingerprint, 63, 167
Fluoroscopy, 37, 81
Food irradiation, 34, 35
Forbidden energy gap, 92, 95, 98, 106, 127
Frenkel defect, 105, 106, 118
Fricke dosimeter, 61
FWHM, 52, 135

Gain, 51, 57
Gamma (γ) rays, 6, 13, 25, 33, 53, 56–58, 60, 63, 67, 71, 73, 77, 81, 112
Gas filled detector, 51, 56
Gate voltage, 70, 71, 150
Geiger-Mueller (GM)
  counter, 53, 58, 82
General effective media (GEM), 139
Glow curve, 62, 63, 68, 69
Gray (Gy), 23, 24
Growth method, 81

Hall
  coefficient, 120
  effect, 117, 120, 153
Heavy charged particles, 72, 77
Homeland security, 164, 189
Hopping conduction, 95, 102, 103

Ideality factor η, 129
impedance spectroscopy, 117
impurity band, 95, 102
In_2O_3/SiO
  thick film, 142, 159
  thin film, 124, 125
In vivo dosimetry, 38, 39
Interdigitated electrode, 6, 138, 153, 179

Interface circuit, 167
Ionization, 5, 11, 12, 14, 15, 18, 46, 47, 56, 61, 62, 75, 91, 104–6, 108, 112
  chamber, 51–53, 56, 71–73
  energy, 52, 75
  pen-sized ionization chamber, 53

Junction, 6, 72, 75–77, 117, 128–31, 139–42, 144, 145, 153

Kyshtym, 40, 42

Lattice imperfection, 46, 95, 116
Leakage current, 6, 72, 74, 128, 131, 140, 142
Light emitting diodes, 75, 185
Linear absorption coefficient μ, 33
  discriminant analysis, 175
  energy transfer (LET), 14, 18, 21–23, 28, 29
  nonthreshold hypothesis, 26, 31
Lithium fluoride (LiF), 17, 18, 20, 63, 64, 67–69
  iodide (LiI), 57, 83
Localized states, 92, 96–98, 102
Lower limit of detection (LLD), 47
Luminescence, 4, 47, 57, 58, 62, 64, 65, 67, 73, 164

Maxwell-Wagner equation, 139
Mean free path, 115
Medical devices sterilization, 2, 3, 33–35, 50
  use of ionizing radiation, 37
Melting point, 36, 117, 122, 161
Metal oxide, 5, 6, 39, 83, 106, 115, 118, 119, 137, 153, 159, 191
  mixture, 119, 122
  semiconductor (MOS), 6, 70–72, 150, 151, 153, 173, 175
Metal-semiconductor contact
  blocking, 100
  neutral contact, 99
  ohmic, 99
Metal-substituted phthalocyanine, 7, 83, 132, 145, 147, 149, 153
Minimum measurable dose. *See* LLD, 68
Mixed transition region, 132, 140
MnO, 6, 122–24
MnO/TeO$_2$ thin film, 122–24
MnPc, 147–50
Mobility, 80, 94, 95, 103, 112, 119, 124

MOS
  capacitor, 150, 151
  dosimeter, 150
MOSFET, 39, 72
Mott and Davis theory, 96–98, 120, 127
multilayer interconnect circuit, 116

Neutron, 1, 2, 11, 12, 14, 23, 25, 30, 49, 50, 56–60, 63, 66, 67, 77, 81–83, 106, 107, 112, 120, 165–67
  detector, 56
NiO, 6, 137–40
NiPc thick film, 6, 146, 147, 149, 150
Normalization, 170
normalized current, 122, 124, 130–32, 142–45, 160, 161, 170
Nuclear accident, 40, 41, 46, 189
  collision, 72, 73
  waste, 3, 39, 46, 84, 186, 189

Olfactory, 163, 172
Optical absorption analysis, 91
  absorption coefficient $\alpha(\nu)$, 64, 92, 96, 97, 135, 138
  band gap, 93, 97, 120, 121, 127, 150
  density, 61, 66, 125, 127, 137, 145, 147, 149
  fiber sensor, 83, 84, 118
  transition, 93, 98
Optically stimulated luminescence (OSL), 4, 65, 67, 68, 84
Oxidation, 5, 61, 108, 111, 112, 118, 120, 132, 140, 141
Oxygen vacancy, 6, 106, 107, 118, 120, 124

Particle range, 15
Pattern recognition, 159, 165, 168
Passivated ion-implanted silicon, 73
Passivation, 74, 117
Percolation, 138, 139
Perovskite, 58
Personal electronic dosimeter, 46, 189
Phosphor, 49, 62, 63, 68
Photocathode, 57
Photodiode, 59, 60, 75, 77
Photoelectric interaction, 74
Photographic emulsion, 65
Photon, 2, 12–14, 24, 30, 33, 49–53, 57–59, 62, 63, 65, 68, 72–74, 80, 92–94, 96, 97, 103, 112, 136

Photon (continued)
  beam attenuation, 73
Phthalocyanine, 6, 7, 83, 115, 132, 147, 150, 153, 162, 191
Pixellated detector, 81
Planar structure, 6, 138, 153
Point defect, 106, 118
Polarizability, 102, 124
Polymer gel, 79
Polymerization, 36, 61, 79
Poole-Frenkel, 102, 104, 134, 135
Potential barrier, 99, 101, 104, 134
Preprocessing, 169, 172
Principal component analysis (PCA), 172–75
Proportional counter, 51, 56, 73
Proton, 1, 11, 12, 14, 15, 18, 22, 30, 56, 58, 71, 77, 105, 112
PVDF, 110, 161, 162

Quality factor Q, 23, 24, 29

rad, 5, 23, 24
RADFET, 70, 71, 191
Radiation
  biological effects, 25
  damage, 5, 25, 26, 27, 31, 77, 105–8, 117, 141, 144, 150, 153
  deterministic, 26, 31
  dose, acute, 26, 27
  dose, chronic, 26, 27
  hardness, 77, 137
  nonionizing, 11
  nose, 7, 8, 159, 163, 165–68, 176, 178
  protection, 2, 3, 22, 25, 31, 32, 48–50
  stochastic, 25, 26, 31
  therapy, 29, 38, 48–50, 72, 79
Radiochromic film, 61, 62
Radiophotoluminescent glass (RPL), 67
Radiotherapy, 38, 39, 79, 80, 84
Raman spectroscopy, 117, 142, 153
Recombination, 65, 76, 105, 107, 121, 128, 129, 131
Refractive index, 97
Relative biological effectiveness (RBE), 28–31
rem, 24, 70, 79
Remote dosimeter, 84
Resistive sensor, 8, 176
Resistivity, 47, 74, 77, 78, 81, 99, 120, 121
Response characteristics, 136, 138, 144, 153

Richardson-Schottky relation, 104
Risk R_T, 29
Roentgen (R), 24
Rutherford, 12, 72

Saturation, 52, 64, 72, 102, 107, 108, 142, 144, 147, 149, 171
Scaling, 171
Scintillation counter, 51, 57, 58, 73
Scintillator, 49, 57–60, 72, 83
Screen-printing, 116
Sellafield, 40, 41
SEM, 117, 142, 153
Semiconductor detector, 71–73, 106
Sensitivity, 3, 7, 47, 49, 51–53, 55–57, 63, 68, 138, 150, 153, 159, 161, 162
Shielding, 32, 33, 63, 68, 73, 137, 163, 166, 168
Short-range order, 92, 95, 127
Schottky
  conduction mechanism, 104
  field-lowering coefficient, 104
Sheet
  density, 121, 122
  resistance, 120
Sievert (Sv), 23, 24
Signal conditioning, 137, 153, 167, 168
Silicon
  diode detector, 72
  surface barrier, 73
  wafer, 128
Smakula, 64, 135
Software, 4, 8, 55, 176, 178, 181, 183, 185
Solid-state ionization, 62, 71, 72
Space-charge-limited conduction, 102, 134, 135
Sterilization, 2, 3, 27, 33–36, 50, 109
Stoichiometry, 124, 141
Stopping power, 15–22, 58, 80, 84
Structural
  defect, 6, 101, 107, 116, 121, 149
  order, 92, 94, 124, 147
Syndrome, acute radiation, 26
  central nervous system, 27
  gastrointestinal tract, 27, 43

Tail states, 95
Tauc, 96, 97
TeO_2 thin film, 119–21
TeO_2/In_2O_3 thin film, 122

TeO$_2$/S, 128
TeO$_2$/Si, 129–31
Thallium-doped sodium iodide NaI(Tl), 57
Thermal
  evaporation, 115
  equilibrium, 76, 99, 100, 105, 121, 128
  vibration, 95
Thermoluminescence, 62, 64
  dosimeter (TLD), 4, 27, 48, 62, 63, 65–69
Thick film technology, 116
Three Mile Island, 39, 40, 43
Threshold voltage, 49, 56, 70–72, 150–52
Tissue-equivalent, 15, 17, 18, 20, 23, 30, 63, 68, 80
Transition
  direct allowed, 97, 120, 127
  direct forbidden, 97
  indirect allowed, 97, 127
  indirect forbidden, 97
Transmittance, 96
Trapping level, 76, 92, 128

Tunneling, 95, 104, 129, 131, 138
Types of disorder, 95

Ultraviolet light (UV), 47, 67
Urbach, 96, 127

Valence band, 5, 76, 91–96, 98, 102, 103, 121, 128, 131
Validation, 8, 34, 172, 175, 176

Weighted dose equivalents in the organs or tissue H_T, 29
Weighting factor, 29, 30
Work function, 99

X-ray diffraction (XRD), 117, 124, 125, 153, 161
  radiation, 2, 11–14, 25, 28, 32–34, 37, 38, 48, 49, 53, 55, 56, 59, 67, 68, 71, 73, 77, 78, 80, 81, 107, 112

Zinc sulfide (ZnS), 57, 72, 73

Related Titles from Artech House

Chemical and Biochemical Sensing with Optical Fibers and Waveguides, Gilbert Boisde and Alan Harmer

Introduction to Microelectromechanical Microwave Systems, Second Edition, Hector J. De Los Santos

Introduction to Microelectromechanical Systems Engineering, Second Edition, Nadim Maluf and Kirt Williams

Measurement Systems and Sensors, Waldemar Nawrocki

MEMS Mechanical Sensors, Steve P. Beeby, Graham Ensel, and Neil M. White

Optical Fiber Sensors, Volume III: Components and Subsystems, Biran Culshaw and John Dakin, editors

Optical Fiber Sensors, Volume IV: Applications, Analysis, and Future Trends, Brian Culshaw and John Dakin, editors

Sensor Technologies and Data Requirements for ITS, Lawrence A. Klein

Understanding Smart Sensors, Second Edition, Randy Frank

For further information on these and other Artech House titles, including previously considered out-of-print books now available through our In-Print-Forever® (IPF®) program, contact:

Artech House
685 Canton Street
Norwood, MA 02062
Phone: 781-769-9750
Fax: 781-769-6334
e-mail: artech@artechhouse.com

Artech House
46 Gillingham Street
London SW1V 1AH UK
Phone: +44 (0)20 7596-8750
Fax: +44 (0)20 7630-0166
e-mail: artech-uk@artechhouse.com

Find us on the World Wide Web at:
www.artechhouse.com